United States Nuclear Regulatory Commission

Protecting People and the Environment

NUREG-1929, Vol. 2

I0482655

Safety Evaluation Report

Related to the License Renewal of Beaver Valley Power Station, Units 1 and 2

Docket Nos. 50-334 and 50-412

FirstEnergy Nuclear Operating Company

Office of Nuclear Reactor Regulation

AVAILABILITY OF REFERENCE MATERIALS
IN NRC PUBLICATIONS

United States Nuclear Regulatory Commission

Protecting People and the Environment

NUREG-1929, Vol. 2

Safety Evaluation Report

Related to the License Renewal of Beaver Valley Power Station, Units 1 and 2

Docket Nos. 50-334 and 50-412

FirstEnergy Nuclear Operating Company

Manuscript Completed: October 2009
Date Published: October 2009

Office of Nuclear Reactor Regulation

ABSTRACT

This safety evaluation report (SER) documents the technical review of the Beaver Valley Power Station (BVPS), Units 1 and 2, license renewal application (LRA) by the United States (US) Nuclear Regulatory Commission (NRC) staff (the staff). By letter dated August 27, 2007, FirstEnergy Nuclear Operating Company (FENOC or the applicant) submitted the LRA in accordance with Title 10, Part 54, of the *Code of Federal Regulations*, "Requirements for Renewal of Operating Licenses for Nuclear Power Plants." FENOC requests renewal of the Units 1 and 2, operating licenses (Facility Operating License Numbers DPR-66 and NPF-73, respectively) for a period of 20 years beyond the current expirations at midnight January 29, 2016, for Unit 1, and at midnight May 27, 2027, for Unit 2.

BVPS is located approximately 17 miles west of McCandless, PA. The NRC issued the construction permits for Unit 1 on June 26, 1970, and on May 3, 1974, for Unit 2. The NRC issued the operating licenses for Unit 1 on July 2, 1976, and on August 14, 1987, for Unit 2. Units 1 and 2 are of a dry subatmospheric pressurized water reactor design. Westinghouse Electric supplied the nuclear steam supply system and Stone and Webster originally designed and constructed the balance of the plant. The licensed power output of each unit is 2900 megawatt thermal with a gross electrical output of approximately 972 megawatt electric.

This SER presents the status of the staff's review of information submitted through June 04, 2009, the cutoff date for consideration in the SER. Section 6.0 provides the staff's final conclusion on the review of the BVPS LRA.

TABLE OF CONTENTS

List of Tables

ABBREVIATIONS

AC	alternating current
ACI	American Concrete Institute
ACU	air conditioning unit
ACRS	Advisory Committee on Reactor Safeguards
ADAMS	Agencywide Document Access and Management System
AEM	aging effect/mechanism
AERM	aging effect requiring management
AFW	auxiliary feedwater
AISC	American Institute of Steel Construction
AMP	aging management program
AMR	aging management review
AMSAC	ATWS Mitigation System Actuation Circuitry
ANSI	American National Standards Institute
ART	adjusted reference temperature
ASME	American Society of Mechanical Engineers
ASTM	American Society for Testing and Materials
ATWS	anticipated transient without scram
BTP	branch technical position
BVPS	Beaver Valley Power Station
BWR	boiling water reactor
CASS	cast austenitic stainless steel
CAT	chemical addition tank
CDF	core damage frequency
CE	electrical continuity
CF	chemistry factor
CFR	*Code of Federal Regulations*
CI	confirmatory item
CLB	current licensing basis
CMAA	Crane Manufacturers Association of America
CO_2	carbon dioxide
CR-15	Unit 1 fuel cask crane
CR-27	Unit 1 moveable platform and hoists crane
CRDM	control rod drive mechanism
CREVS	control room emergency ventilation system
CRN201	Unit 2 polar crane
CRN215	Unit 2 spent fuel cask trolley
CRN227	Unit 2 moveable platform with hoists
CUF	cumulative usage factor
C_vUSE	Charpy upper shelf energy
DBA	design basis accident
DBE	design basis event
DC	direct current
DF	direct flow

DLCo	Duquesne Light Company
ECCS	emergency core cooling system
EDG	emergency diesel generator
EFPY	effective full-power year
El	elevation
EN	enclosure or protection
EOL	end-of-license (current license life)
EOLE	end-of-license-extended (end of renewed license life)
EPRI	Electric Power Research Institute
EQ	environmental qualification
ER	applicant's environmental report
ERF	emergency response facility
ESF	engineered safety features
EXP	expansion or separation
FAC	flow-accelerated corrosion
FB	fire barrier
F_{en}	environmental fatigue life correction factor
FLB	flood barrier
FP	fire protection
FR	*Federal Register*
FSAR	final safety analysis report
ft-lb	foot-pound
FW	feedwater
GALL	Generic Aging Lessons Learned Report
GDC	general design criteria or general design criterion
GEIS	Generic Environmental Impact Statement
GL	generic letter
GSI	generic safety issue
HELB	high-energy line break
HHSI	high head safety injection
HLBS	HELB shielding
HS	heatsink
HVAC	heating, ventilation, and air conditioning
HX	heat exchanger
I&C	instrumentation and controls
IASCC	irradiation assisted stress corrosion cracking
IEB	inspection and enforcement bulletin
IEEE	Institute of Electrical and Electronics Engineers
IGA	intergranular attack
IGSCC	intergranular stress corrosion cracking
IN	information notice
INE	insulate (electrical)
INPO	Institute of Nuclear Power Operations
IPA	integrated plant assessment
ISG	interim staff guidance

ISI	inservice inspection
ksi	1000 pounds (kilo-pound) per square inch
kV	kilo-volt
LBB	leak-before-break
LER	licensee event report
LHSI	low head safety injection
LOCA	loss of coolant accident
LR	license renewal
LRA	license renewal application
LTOP	low-temperature overpressure protection
MB	missile barrier
MIC	microbiologically-influenced corrosion
MWe	megawatts-electric
MWt	megawatts-thermal
N_2	nitrogen
NA	not applicable
NaOH	sodium hydroxide
n/cm^2	neutrons per square centimeter
NDE	nondestructive examination
NDT	nil-ductility transition
NEI	Nuclear Energy Institute
NFPA	National Fire Protection Association
NRC	US Nuclear Regulatory Commission
NSSS	nuclear steam supply system
ODSCC	outside-diameter stress corrosion cracking
OI	open item
OPPS	overpressure protection system
pH	potential hydrogen
PMF	probable maximum flood
ppm	parts per million
PR	pressure relief
P-T	pressure-temperature
PTS	pressurized thermal shock
PVC	polyvinyl chloride
PW	pipe whip restraint
PWR	pressurized water reactor
PWSCC	primary water stress corrosion cracking
QA	quality assurance
RAI	request for additional information
RCCA	rod cluster control assembly
RCPB	reactor coolant pressure boundary
RCS	reactor coolant system

RG	regulatory guide
RHR	residual heat removal
RIS	regulatory issue summary
RP	gaseous relief path
rpm	revolutions per minute
RPV	reactor pressure vessel
RT	reference temperature
RT_{NDT}	reference temperature for nil ductility transition
ΔRT_{NDT}	shift in reference temperature for nil ductility transition
RT_{PTS}	reference temperature for pressurized thermal shock
RVI	reactor vessel internals
RWST	refueling water storage tank
SBO	station blackout
SC	structure and component
SCC	stress-corrosion cracking
SCW	shutdown cooling water
SER	safety evaluation report
SG	steam generator
SHD	shielding
SIS	safety injection system
SNS	support for Criterion (a)(2) equipment
SPB	structural pressure barrier
SOC	statement of consideration
SRE	support for Criterion (a)(3) equipment
SRP	Standard Review Plan
SRP-LR	Standard Review Plan for Review of License Renewal Applications for Nuclear Power Plants
SSC	system, structure, and component
SSE	safe-shutdown earthquake
SSR	support for Criterion (a)(1) equipment
t/4	one fourth of the way through the vessel wall
TLAA	time-limited aging analysis
TS	technical specifications
U_{60}	sixty year cumulative usage factor
U_{env}	cumulative usage factor which includes environmental effects
UFSAR	Updated Final Safety Analysis Report
USE	upper-shelf energy
UT	ultrasonic testing
UV	ultraviolet
VAC	volts alternating current
WANO	World Association of Nuclear Operators
WASS	wrought austenitic stainless steel
WCAP	Westinghouse Commercial Atomic Power
Zn	zinc

SECTION 4

TIME-LIMITED AGING ANALYSES

4.1 Identification of Time-Limited Aging Analyses

This Section of the safety evaluation report (SER) addresses the identification of time-limited aging analyses (TLAAs). In license renewal application (LRA) Sections 4.2 through 4.7, FirstEnergy Nuclear Operating Company (FENOC or the applicant) addressed the TLAAs for Beaver Valley Power Station Units 1 and 2. SER Sections 4.2 through 4.8 documents the review of the TLAAs conducted by the staff of the United States (US) Nuclear Regulatory Commission (NRC) (the staff).

TLAAs are certain plant-specific safety analyses that involve time-limited assumptions defined by the current operating term. Pursuant to Title 10, Section 54.21(c)(1), of the *Code of Federal Regulations* (10 CFR 54.21(c)(1)), applicants must list TLAAs as defined in 10 CFR 54.3.

In addition, pursuant to 10 CFR 54.21(c)(2), applicants must list plant-specific exemptions granted under 10 CFR 50.12 based on TLAAs. For any such exemptions, the applicant must evaluate and justify the continuation of the exemptions for the period of extended operation.

Unit 1 and Unit 2 are constructed of similar materials with similar environments. Therefore, the mechanical system and component information presented in the LRA typically applies to both units, and no unit-specific identifier is listed. However, design differences exist between Unit 1 and Unit 2. Those design differences are identified by using a designator (*i.e.*, Unit 1 only or Unit 2 only). Further, the applicant assigned a different designator (*i.e.*, common) for those cases where the system, structure, or component is used and/or shared by both units.

4.1.1 Summary of Technical Information in the Application

To identify the TLAAs, the applicant evaluated calculations for BVPS against the six criteria specified in 10 CFR 54.3. The applicant indicated that it identified the calculations that met the six criteria by searching the current licensing basis (CLB). The CLB includes the updated final safety analysis report (UFSAR), engineering calculations, technical reports, engineering work requests, licensing correspondence, and applicable vendor reports. In LRA Table 4.1-1, "List of BVPS Time-Limited Aging Analyses and Resolution," the applicant listed the applicable TLAAs:

- reactor vessel neutron embrittlement
- metal fatigue
- environmental qualification (EQ) of electric equipment
- concrete containment tendon prestress
- containment liner plate, metal containment, and penetrations fatigue
- piping subsurface indications (Unit 1 only)
- reactor vessel underclad cracking (Unit 1 only)
- main coolant loop piping leak-before-break
- pressurizer surge line piping leak-before-break
- branch line piping leak-before-break (Unit 2 only)

- high-energy line break postulation
- settlement of structures (Unit 2 only)
- crane load cycles

Pursuant to 10 CFR 54.21(c)(2), the applicant stated that it did not identify exemptions granted under 10 CFR 50.12 based on a TLAA as defined in 10 CFR 54.3.

4.1.2 Staff Evaluation

In LRA Section 4.1, the applicant listed the Unit 1 and Unit 2 TLAAs. The staff reviewed the information to determine whether the applicant has provided sufficient information pursuant to 10 CFR 54.21(c)(1) and 10 CFR 54.21(c)(2).

As defined in 10 CFR 54.3, TLAAs meet the following six criteria:

(1) involve systems, structures, and components within the scope of license renewal, as described in 10 CFR 54.4(a)

(2) consider the effects of aging

(3) involve time-limited assumptions defined by the current operating term (40 years)

(4) are determined to be relevant by the applicant in making a safety determination

(5) involve conclusions, or provide the basis for conclusions, related to the capability of the system, structure, and component to perform its intended functions, as described in 10 CFR 54.4(b)

(6) are contained or incorporated by reference in the CLB

The applicant reviewed the list of common TLAAs in NUREG-1800, Revision 1, "Standard Review Plan for Review of License Renewal Applications for Nuclear Power Plants" (SRP-LR), dated September 2005. The applicant listed TLAAs applicable to BVPS in LRA Table 4.1-1.

As required by 10 CFR 54.21(c)(2), the applicant must list all exemptions granted under 10 CFR 50.12, based on TLAAs, and evaluated and justified for continuation through the period of extended operation. The LRA states that each active exemption was reviewed to determine whether it was based on a TLAA. The applicant did not identify any TLAA-based exemptions. Based on the information provided by the applicant regarding the process used to identify these exemptions and its results, the staff concludes, in accordance with 10 CFR 54.21(c)(2), that there are no TLAA-based exemptions justified for continuation through the period of extended operation.

4.1.3 Conclusion

Based on its review, the staff concludes that the applicant has provided an acceptable list of TLAAs, as required by 10 CFR 54.21(c)(1). The staff confirms, as required by 10 CFR 54.21(c)(2), that no exemption pursuant to 10 CFR 50.12 has been granted based on a TLAA.

4.2 Reactor Vessel Neutron Embrittlement

"Neutron embrittlement" describes changes in the mechanical properties of reactor vessel (RV) materials (or any other ferrous materials) resulting from exposure to fast neutron (E>1.0 MeV) fluence. The rate of neutron exposure is defined as neutron flux, and the cumulative degree of exposure over time is defined as neutron fluence. Fracture toughness of ferritic materials not only depends on fluence but, also on temperature. The area within the vicinity of the reactor core called the beltline region is defined by 10 CFR 50.61 as:

> The region of the reactor vessel (shell material including welds, heat-affected zones and plates and forgings) that directly surrounds the effective height of the active core and adjacent regions of the reactor vessel that are predicted to experience sufficient neutron radiation damage to be considered in the selection of the most limiting material with regard to radiation damage.

The most pronounced material change (and most relevant in this review) is reduction in fracture toughness. Neutron fluence reduces fracture toughness, which is the material's resistance to crack propagation. The reference temperature for nil-ductility transition (RT_{NDT}) is a metric of the temperature above which the material becomes ductile and below which it becomes brittle. RT_{NDT} increases with fluence, meaning higher temperatures are required for the material to remain ductile. In pressure vessel applications, the RT_{NDT} is increased by a margin term added to account for uncertainties from the available limited materials data.

In addition to the beltline region, materials in the extended beltline region above or below the beltline that exceed fluence values of $1.0E+17$ n/cm^2 (E>1.0 MeV) are subject to the requirements of 10 CFR 50 Appendix H and must be monitored for evaluation of changes in fracture toughness at end-of-license-extended.

Determination of the reduction in fracture toughness, affects several RV analyses that support plant operation, including:

- Neutron Fluence Values;
- Pressurized Thermal Shock (PTS);
- Charpy V notch Upper-Shelf Energy CvUSE); and
- Pressure-Temperature (P-T) Limits

These analyses of the reduction of fracture toughness of the RVs for 40 calendar years are TLAAs and must be evaluated for the period of extended operation.

4.2.1 Neutron Fluence Values

4.2.1.1 Summary of Technical Information in the Application

In LRA Section 4.2.1, the applicant summarized its evaluation of neutron fluence values for the period of extended operation as follows:

> The loss of fracture toughness is an aging effect caused by the neutron embrittlement aging mechanism resulting from prolonged exposure to neutron radiation. This process increases tensile strength and hardness of the material

4-3

and reduces toughness. The rate of neutron exposure is defined as neuron flux and the cumulative degree of exposure over time is called neutron fluence. As neutron embrittlement progresses, the toughness/temperature curve shifts downward (lower fracture toughness) and to the right (brittle/ductile transition as temperature increases).

4.2.1.1.1 Unit 1

In the spring of 2000, Surveillance Capsule Y was pulled for analysis and was documented in Westinghouse Commercial Atomic Power (WCAP)-15571, "Analysis of Capsule Y from First Energy Company Beaver Valley Unit 1 Reactor Vessel Radiation Surveillance Program." For license renewal, WCAP-15571, Supplement 1, "Analysis of Capsule Y from First Energy Company Beaver Valley Unit 1 Reactor Vessel Radiation Surveillance Program," documents the end-of-license-extended analysis for neutron fluence values.

LRA Tables 4.2-1 and 4.2-2 show the calculated fast neutron fluence (E> 1.0 MeV) values at the inner surface of the Unit 1 RV for the beltline and extended beltline materials, respectively. These values, projected by Evaluated Nuclear Data File/B-VI cross sections, are based on the results of the Capsule Y analysis and comply with Regulatory Guide (RG) 1.190, "Calculational and Dosimetry Methods for Determining Pressure Vessel Neutron Fluence."

These fluence data tabulations include fuel cycle-specific calculated neutron exposures at the end of Cycle 17 (February 2006) as well as future projections to the end of Cycle 18 and for several intervals extending to 54 effective full-power years (EFPYs). The calculations account for a core power uprate from 2689 megawatts-thermal (MWt) to 2900 MWt at the onset of Cycle 18. Neutron exposure projections beyond the end of Cycle 17 are based on the spatial power distributions and Cycle 18 plant characteristics at the uprated power level.

4.2.1.1.2 Unit 2

In the spring of 2005, Surveillance Capsule X was pulled for analysis and was documented in WCAP-16527-NP, "Analysis of Capsule X from First Energy Nuclear Operating Company Beaver Valley Unit 2 Reactor Vessel Radiation Surveillance Program." For license renewal, WCAP-16527-NP, Supplement 1, "Analysis of Capsule X from First Energy Company Beaver Valley Unit 2 Reactor Vessel Radiation Surveillance Program," documents the end-of-license-extended analysis for neutron fluence values.

LRA Tables 4.2-3 and 4.2-4 show the calculated fast neutron fluence (E> 1.0 MeV), values at the inner surface of the Unit 2 RV for the beltline and extended beltline materials, respectively. These values, projected by Evaluated Nuclear Data File/B-VI cross sections and based on the results of the Capsule X analysis, comply with RG 1.190.

These fluence data tabulations include fuel cycle-specific calculated neutron exposures at the end of Cycle 11 (April 2005) as well as future projections for several intervals extending to 54 EFPYs based on assumptions that the core power distributions and plant operating characteristics for Cycle 12 represent plant operation to 17 EFPYs and that the preliminary Cycle 13 core power distributions apply beyond 17 EFPYs. The calculations account for a core power uprate from 2689 to 2900 MWt at 17 EFPYs.

4.2.1.2 Staff Evaluation

The staff reviewed LRA Section 4.2.1, pursuant to 10 CFR 54.21(c)(1)(i)(ii) or (iii).

4.2.1.2.1 Unit 1

The applicant submitted an updated Surveillance Capsule Y analysis report in WCAP-15571-NP, Supplement 1. Capsule Y was removed in the spring of 2000 and Supplement 1 was prepared for license renewal.

In the LRA, the applicant stated that the methodology used for the calculation of the applicable fluence values adheres to the guidance in RG 1.190 therefore, it is acceptable. The applicant further stated that it had investigated the materials which extend above and below the beltline region where fluence values were above 1×10^{17} n/cm^2, as specified in 10 CFR 50, Appendix H. Projected values were calculated for 54 EFPYs of operation, accounting for a very conservative 90% load factor for the entire period of operation. This calculation accounted for a power uprate from 2689 MWt to 2900 MWt. To estimate the neutron source, the applicant assumed that the current cycle (Cycle 18) loading will be the average cycle loading to the end of the extended license.

For Unit 1, the critical element regarding the reference temperature for pressurized thermal shock (RT_{PTS}) is the lower shell plate B6903-1. The peak inside surface fluence value for 54 EFPYs is 6.09×10^{19} n/cm^2. However, at 54 EFPYs, the RT_{PTS} value is 275.7 ˚F, which exceeds the screening criterion of 270 ˚F pursuant to 10 CFR 50.61. The applicant estimated (and the staff verified) that the RT_{PTS} screening criterion will be reached at 4.96×10^{19} n/cm^2 or, 43.87 EFPYs. Therefore, the validity of RT_{PTS}, P-T limit curves and LTOP limits (and the associated technical specification (TS) limits) will be valid to 43.87 EFPYs. The staff determined that additional information was required to complete its review. In a request for additional information (RAI) 4.2.4-1, dated April 28, 2008, the staff requested that the applicant provide additional details or documentation to show whether or not the low-temperature overpressure protection system was affected by the extended power uprate.

In its response to RAI 4.2.4-1, dated May 28, 2008, the applicant stated that LTOP set-points were calculated in WCAP-16799-NP and that the analyses include the power uprate to 2900 MWt.

Based on its review, the staff finds the applicant's response to RAI 4.2.4-1 for Unit 1 acceptable because the applicant's revised LTOP fluence analysis used staff approved methods and the set-points accounted for the power uprate. Therefore, the staff's concern described in RAI 4.2.4-1 for BVPS 1 is resolved.

4.2.1.2.2 Unit 2

The applicant provided Surveillance Capsule X analysis report WCAP-16527-NP, Supplement 1. Capsule X was removed in March 2005 and Supplement 1 was prepared for the license extension application.

In the LRA, that applicant stated that the methodology it had used adheres to the guidance in RG 1.190, therefore, it is acceptable. The materials investigated extended above and below the beltline region where fluence values were above 1×10^{17} n/cm^2 as specified in 10 CFR 50 Appendix H. Projected values are calculated for 54 EFPYs of operation that accounts for a 90% load factor for the entire period of operation that is very conservative. The calculation accounted for power uprate from 2689 MWt to 2900 MWt. To estimate the neutron source it is assumed that the Cycle 13 loading (17 EFPYs and beyond) will be the average cycle loading to the end of the extended license.

For Unit 2, the critical element in the belt region in intermediate shell plate B9004-1 and in the extended beltline region is the upper shell plate B9003-2. The peak fluence calculated for the intermediate shell plate is 6.22×10^{19} n/cm^2 and for the upper shell plate is 0.492×10^{19} n/cm^2. The resulting RT$_{PTS}$ values of 152.4 ˚F and 160.6 ˚F respectively, are lower than the 10 CFR 50.61 screening requirement of 270 ˚F; therefore, these values are acceptable for 54 EFPYs. The staff determined that additional information was required to complete its review.

In RAI 4.2.4-1 dated April 28, 2008, the staff requested that the applicant provide additional details or documentation to show whether or not the LTOP system was affected by the extended power uprate.

In its response to RAI 4.2.4-1, dated May 28, 2008, the applicant stated that the LTOP set-points have been calculated in WCAP-15677, reflected in the Unit 2 P-T limits report (PTLR), and is valid for 22 EFPYs.

Based on its review, the staff finds the applicant's response to RAI 4.2.4-1 for Unit 2 acceptable because the applicant's revised LTOP fluence analysis used staff approved methods and the set-points accounted for the power uprate. Therefore, the staff's concern described in RAI 4.2.4-1 for Unit 2 is resolved.

4.2.1.3 UFSAR Supplement

The applicant provided a UFSAR supplement summary description of its TLAA evaluation of neutron fluence values in LRA Sections A.2.2.1 and A.3.2.1. Based on its review of the UFSAR supplement, the staff concludes that the summary description of the applicant's actions to address neutron fluence values is adequate.

4.2.1.4 Conclusion

Based on its review, as discussed above, the staff concludes that the applicant has demonstrated that the proposed fluence values to the end-of-license-extended are acceptable because they adhere to the guidance of RG 1.190. The staff also concludes that the UFSAR supplement contains an appropriate summary description of the TLAA evaluation, as required by 10 CFR 54.21(d).

4.2.2 Pressurized Thermal Shock

4.2.2.1 Summary of Technical Information in the Application

In LRA Section 4.2.2, the applicant summarized the evaluation of PTS for the period of extended operation. For protection against PTS events for pressurized-water reactors (PWRs), 10 CFR 50.61(b)(1) requires licensees to update assessments of reference temperature projected values whenever a significant change occurs in projected values for adjusted RT_{PTS} or upon a request for a change in the expiration date for the operation of the facility. Irradiation by high-energy neutrons raises the RT_{NDT} value for the RV. Determination of the initial RT_{NDT} is through testing of unirradiated material specimens. The shift in reference temperature, ΔRT_{NDT}, is the difference in the 30 ft-lb index temperatures from the average Charpy curves measured before and after irradiation. RG 1.99, Revision 2, "Radiation Embrittlement of Reactor Vessel Materials," defines the calculation methods for ΔRT_{NDT} and end-of-license upper shelf energy (USE). Determination of RT_{PTS}, defined as the RT_{NDT} value evaluated at the end-of-license fluence for each of the vessel beltline materials, is by two methods pursuant to 10 CFR 50.61(c) and described in RG 1.99, Revision 2, as Regulatory Positions (RPs) 1 and 2. RP 1 applies for material with no credible surveillance data available; RP 2, when credible material surveillance data is available. Adjusted reference temperature (ART) calculations for both RP 1 and 2 follow the guidance in RG 1.99, Revision 2, Sections 1.1 and 2.1, respectively, using copper and nickel content of beltline materials and end-of-license best estimate fluence projections. RT_{PTS} screening criteria established pursuant to 10 CFR 50.61(b)(2) are 270 °F for plates, forgings, and axial welds and 300 °F for circumferential welds.

4.2.2.1.1 Unit 1

Actions to manage the RV fluence at the limiting location have been underway at Unit 1 since the 1990s. Starting with Cycle 11 in 1995, BVPS instituted a flux management program for the fluence effects on the RT_{PTS} value of the limiting plate (lower shell plate B6903-1).

This program added hafnium rods in the peripheral fuel bundles and continued use of the standard L4P low-leakage core loading. The applicant submitted an updated RT_{PTS} analysis demonstrating that the limiting beltline plate would meet 10 CFR 50.61 requirements at the end-of-license fluence with no further flux management initiatives.

In the SER issued October 7, 1997, addressing the PTS status for Unit 1, the staff agreed and determined that the RT_{PTS} value for the limiting beltline component (plate B6903-1) at the end of the current operating term would be 267.8 °F and that BVPS 1 met 10 CFR 50.61 requirements. The BVPS 1 operation with hafnium rods installed for three cycles (removed in fall of 2001) reduced the irradiation rate by approximately 25 percent during that time period. Using the calculated chemistry factor and fluence values of the 1997 SER, the applicant determined that BVPS 1 PTS projections would remain below PTS screening criteria through the end-of-license.

In the spring of 2000, Surveillance Capsule Y was pulled for analysis documented in WCAP-15571. For license renewal, WCAP-15571, Supplement 1, documents the end-of-license-extended analysis for PTS.

Using the prescribed PTS Rule (10 CFR 50.61) methodology, the applicant generated RT_{PTS} values for beltline and extended beltline region materials of the BVPS 1 RV for fluence values at end-of-license-extended (54 EFPYs). The data for the surveillance program plate material were not credible; therefore, the applicant used the data with a σ_Δ (standard deviation for ΔRT_{NDT}) margin of 17 °F. The data for the BVPS 1 surveillance program weld material were credible; therefore, the applicant used a σ_Δ margin of 14 °F. The surveillance capsule materials are representative of the actual vessel plates and intermediate shell longitudinal weld. Chemistry factor values for the BVPS 1 beltline region materials were based on RG 1.99, Revision 2, RPs 1.1 and 2.1, for the BVPS 1 extended beltline materials on RP 1.1.

LRA Table 4.2-5 shows the RT_{PTS} values at 54 EFPYs for the BVPS 1 beltline materials. Evaluation of extended beltline materials likely to receive fluence values greater than 1.0E+17 n/cm^2 (E>1.0 MeV) determined that none of these materials were limiting. The projected RT_{PTS} values for end-of-license-extended (54 EFPYs) meet 10 CFR 50.61 screening criteria for beltline and extended beltline materials except for lower shell plate B6903-1 (heat C6317-1). The 275.7 °F RT_{PTS} for lower shell plate B6903-1 slightly exceeds the criteria. The 270 °F screening limit for lower shell plate B6903-1 will be reached at a fluence level of 4.961E+19 n/cm^2 (E>1.0 MeV), equivalent to 43.87 EFPYs. By projection, the BVPS 1 RV will reach the PTS screening criterion of 270 °F on the limiting plate (B6903-1) in the year 2033.

Section 50.61 of Title 10 of the Code of Federal Regulations allows that:

> For each pressurized water nuclear power reactor for which the value of RTPTS for any material in the beltline is projected to exceed the PTS screening criterion using the EOL fluence, the licensee shall implement those flux reduction programs that are reasonably practicable to avoid exceeding the PTS screening criterion set forth in Paragraph (b)(2) of this section.

Therefore, a sensitivity assessment of available flux reduction measures included several fuel management scenarios (e.g., low-leakage core design, low-power peripheral fuel assemblies, reinsertion of hafnium rods, and the use of part-length shielded assemblies) and several assumed capacity factors up to 98 percent.

Several flux reduction options are available to maintain the limiting plate below the PTS screening criterion to the end-of-license-extended. The Reactor Vessel Integrity Program will manage flux reduction. Documentation of a flux reduction program for BVPS 1 will be in accordance with 10 CFR 50.61.

Monitoring of the BVPS 1 RV fluence will continue under the Reactor Vessel Integrity Program to keep the projected fluence below that assumed for the relevant neutron embrittlement TLAA; therefore, management of the BVPS 1 RT_{PTS} TLAA will be adequate for the period of extended operation in accordance with 10 CFR 54.21(c)(1)(iii).

4.2.2.1.2 Unit 2

In the spring of 2005, Surveillance Capsule X was pulled for analysis documented in WCAP-16527-NP. For license renewal, WCAP-16527-NP Supplement 1 documents the end-of-license-extended analysis for PTS.

Using the prescribed PTS Rule (10 CFR 50.61) methodology, the applicant generated RT_{PTS} values for beltline and extended beltline region materials of the Unit 2 RV for fluence values at end-of-license-extended (54 EFPYs). The data for the surveillance program plate material are credible; therefore, the applicant used a σ_Δ margin of 8.5 °F. The data for the Unit 2 surveillance program weld material are credible; therefore, the applicant used a σ_Δ margin of 14°F. The surveillance capsule materials are representative of the actual vessel plate and intermediate shell longitudinal weld. Chemistry factor values for the Unit 2 beltline region materials were based on RG 1.99, Revision 2, RPs 1.1 and 2.1, for the Unit 2 extended beltline materials on RP 1.1.

LRA Table 4.2-6 shows the RT_{PTS} values at 54 EFPYs for the Unit 2 beltline materials. The applicant also evaluated the extended beltline materials likely to receive fluence values greater than 1.0E+17 n/cm^2 (E>1.0 MeV). The limiting plate material is the upper shell plate (B9003-2) with a projected end-of-license-extended RT_{PTS} value of 160.6 °F for 54 EFPYs. The limiting weld material is the upper shell longitudinal weld (heat number BOHB (E-8018)) with an end of-license-extended RT_{PTS} value of 128.8 °F. The projected RT_{PTS} values for end-of-license-extended (54 EFPYs) meet 10 CFR 50.61 screening criteria for beltline and extended beltline materials; therefore, disposition of the Unit 2 RT_{PTS} TLAA is in accordance with 10 CFR 54.21(c)(1)(ii).

4.2.2.2 Staff Evaluation

The staff reviewed LRA Section 4.2.2 to verify, pursuant to 10 CFR 54.21(c)(1)(ii), that the analyses have been projected to the end of the period of extended operation and, pursuant to 10 CFR 54.21(c)(1)(iii), that the effects of aging on the intended function(s) will be adequately managed for the period of extended operation.

The PTS evaluation provides a means for assessing the susceptibility of the RV beltline materials to PTS events in order to ensure that these materials have adequate fracture toughness to support reactor operation, pursuant to the methods of evaluation and safety criteria of 10 CFR 50.61. The staff's review covered the applicants PTS methodology and RT_{PTS} calculations at the end of the period of extended operation, taking into consideration the effects of neutron embrittlement. The acceptance criteria for PTS are based on (1) 10 CFR 50, Appendix A, General Design Criterion (GDC)-14, which requires that the reactor coolant pressure boundary (RCPB) be designed, fabricated, erected, and tested so as to have an extremely low probability of abnormal leakage, of rapidly propagating failure, and of gross rupture; (2) GDC-31, which requires that the RCPB be designed with margin sufficient to assure that, under specified conditions, it will behave in a nonbrittle manner and minimize the probability of a rapidly propagating fracture; and (3) 10 CFR 50.61, which sets fracture toughness criteria for protection against PTS events.

The requirements in 10 CFR 50.61 are established to protect PWR vessels against the consequences of PTS events. The rule requires licensees operating PWRs to calculate end-of-license RT_{PTS} values (as defined in 10 CFR 50.61) for each base metal and weld material in the RV constructed from carbon or low alloy steel materials. The rule also requires that RT_{PTS} values remain below the PTS screening criteria throughout the serviceable life of the facilities. The rule sets a maximum limit of 270 °F for RT_{PTS} values that are calculated for base metals (i.e., forging and plate materials) and axial weld materials and a maximum limit of 300 °F for RT_{PTS} values that are calculated for circumferential weld materials.

Section 50.61 of 10 CFR provides requirements for calculating these RT_{PTS} values, similar to the calculation methodology described in RG 1.99, Revision 2, for determining ART values. 10 CFR 50.61 requires that these calculations account for the effects of neutron radiation and incorporate any relevant RV surveillance capsule data required for reporting as part of the licensee's implementation of its RV materials surveillance program. 10 CFR 50.61 defines RT_{PTS} as the RT_{NDT} value at the clad/base metal interface evaluated for the end-of-license fluence. Therefore, RT_{PTS} is equal to the sum of the initial RT_{NDT}, ΔRT_{NDT}, and the margin term. ΔRT_{NDT} is the product of a chemistry factor and a fluence factor. The chemistry factor is dependent upon the amount of copper and nickel in the material and may be determined using tables provided in 10 CFR 50.61 and is equivalent to the method described in RG 1.99, Revision 2, RP 1.1, for ART calculations. The chemistry factor also may be determined from credible surveillance data and is equivalent to the method described in RG 1.99, Revision 2, RP 2.1, for ART calculations. If credible surveillance data is available for the RV beltline material being analyzed, then the surveillance data must be used when it results in an RT_{PTS} value that is higher (*i.e.*, more conservative) than was calculated using the tables. Either method may be used if the credible surveillance data results in a lower RT_{PTS} value. The fluence factor is dependent upon the neutron fluence at end-of-license. The margin term is defined in 10 CFR 50.61. In accordance with 10 CFR 50.61, ΔRT_{NDT} is a function of neutron fluence. Since neutron fluence changes with time, the determination of ΔRT_{NDT} (and, therefore, RT_{PTS}) meets the TLAA criteria of 10 CFR 54.3(a).

In LRA Section 4.2.2, the applicant discussed its analysis of the Unit 1 and Unit 2 PTS for the period of extended operation. The applicant noted that the Unit 1 PTS analysis includes surveillance data from the analysis of Capsule Y, which was pulled from the Unit 1 RV in the spring of 2000. Likewise, the applicant noted that the PTS analysis for Unit 2 includes surveillance data from the analysis of Capsule X, which was pulled from the Unit 2 RV in the spring of 2005. The latest surveillance capsule reports provided by the applicant are WCAP-15571, "Analysis of Capsule Y from Beaver Valley Unit 1 Reactor Vessel Radiation Surveillance Program," Revision 0, November 2000, and WCAP-16527-NP, "Analysis of Capsule X from First Energy Nuclear Operating Company Beaver Valley Unit 2 Reactor Vessel Radiation Surveillance Program," Revision 0, March 2006. The applicant also referenced WCAP-15571, Supplement 1, July 2007, for Unit 1 and WCAP-16527-NP, Supplement 1, July 2007, for Unit 2.

These supplements contain detailed calculations of the RT_{PTS} and upper shelf energy (USE) values for the end of the period of extended operation (54 EFPYs), based on the latest surveillance data. The applicant stated that the Unit 1 and Unit 2 RT_{PTS} calculations follow the requirements specified in 10 CFR 50.61. The applicant also discussed its RT_{PTS} calculations for the limiting RV beltline material at Unit 1.

The limiting beltline material at Unit 1 is Lower Shell Plate B6903-1 (Heat No. C6317-1). According to the applicant, the RT_{PTS} value for this material at the end of the period of extended operation is 275.7 °F, which exceeds the 270 °F PTS screening limit for plates. The applicant determined that the 270 °F screening limit for this material will be reached at a fluence level of $4.961 \times 10^{19} n/cm^2$ (E > 1.0 MeV). The applicant projected that this fluence level will be reached in the year 2033 (43.87 EFPYs). The applicant stated that the RT_{PTS} values for all other RV beltline materials at Unit 1 and the RT_{PTS} values for all RV beltline materials at Unit 2 are projected to meet the PTS screening criteria of 10 CFR 50.61. Likewise, all extended beltline materials for Unit 1 and Unit 2 will meet the 10 CFR 50.61 screening criteria.

The applicant provided RT_{PTS} values for beltline materials in LRA Tables 4.2-5 and 4.2-6 for Unit 1 and Unit 2, respectively. These tables included all the input data required to determine the RT_{PTS} values at the end of the period of extended operation, including the weight percentage copper and nickel, initial RT_{NDT} values, chemistry factor values, clad/base metal interface fluence values, fluence factors, M values, and margin component terms (σ_i and σ_Δ). The staff independently confirmed that the applicant utilized valid weight percentages for copper and nickel, and initial RT_{NDT} values for the Unit 1 and Unit 2 RV beltline materials. For all RV beltline materials represented in the Unit 1 and Unit 2 RV surveillance programs, as well as those represented in sister plant surveillance programs, the applicant provided two sets of RT_{PTS} calculations. The first calculation was based on the use of the chemistry factor tables from 10 CFR 50.61 (hereafter designated as RG 1.99, Revision 2, RP 1.1 – the equivalent ART methodology). The second calculation was based on the application of the surveillance data (hereafter designated as RG 1.99, Revision 2, RP 2.1). Only one of these two methods may be used for calculating licensing basis chemistry, ART, and RT_{PTS} values for each material, and the applicant did not indicate which of the two calculations represented their proposed licensing basis.

In RAI 1, dated March 5, 2008, the staff requested that the applicant indicate which RP (RP 1.1 or RP 2.1) was used to determine the licensing basis chemistry factor, ART, and RT_{PTS} values for each RV beltline material. The staff also requested that the applicant justify its selection of RPs 1.1 or 2.1 for each material, based on factors such as surveillance data credibility or non-credibility, conservatism of RP 2.1 data, or other factors, such as a previous staff recommendation that non-credible surveillance data be used for calculating the chemistry factor for limiting plate B6903-1 with full σ_Δ margin of 17 °F.

In its response dated April 2, 2008, the applicant stated that in all cases where credible surveillance data are available at Unit 1 and Unit 2, the more conservative of the two RPs is used for determining the licensing basis chemistry factor, ART, and RT_{PTS} values for the material. Specifically, the RP 1.1 or 2.1 resulting in the higher set of values (chemistry factor, ART, and RT_{PTS}) is taken to supersede the RP resulting in the lower set of values for each RV beltline material represented. Furthermore, the applicant stated that for the Unit 1 limiting RV beltline material (Lower Shell Plate B6903-1 (Heat No. C6317-1)), the non-credible surveillance data were used with full σ_Δ margin of 17 °F for determining the CF, ART, and RT_{PTS} values.

In this instance, the use of the non-credible surveillance data with full σ_Δ margin of 17 °F results in higher and, therefore, more conservative chemistry factor, ART, and RT_{PTS} values than those which would be obtained using RP 1.1. The staff finds that this is consistent with the February 12, 1998, NRC-Industry meeting where the staff recommended that the applicant, in this instance, utilize the non-credible surveillance data with a full σ_Δ margin of 17 °F for this plate, because it yields appropriately conservative results. This was also confirmed by the applicant's statement in LRA Section 4.2.2 that the RT_{PTS} value for the Unit 1 limiting RV beltline material, Lower Shell Plate B6903-1, is 275.7 °F at the end of the period of extended operation.

Based on its review, the staff concludes that for all Unit 1 and Unit 2 RV beltline materials, the applicant has utilized the RP that yields the most appropriate results. The staff finds the applicant's response to RAI 1 acceptable because the applicant provided the necessary information regarding their implementation of RP 1.1 or 2.1 for determining the licensing basis chemistry factor, ART, and RT_{PTS} values for the RV beltline materials at Unit 1 and Unit 2 and justified the use of the RP methodology based on maintaining adequate conservatism for these calculations. Therefore, the staff's concern described in RAI 1 is resolved.

For Unit 1, the staff noted several discrepancies between LRA Section 4.2.2, LRA Table 4.2-5, Appendix D of WCAP-15571 (Surveillance Data Credibility Analysis), and WCAP-15571, Supplement 1, regarding the application of surveillance data for determining the RT_{PTS} value for

Intermediate Shell Longitudinal Weld 19-714 (Heat 305424). First, WCAP-15571, Appendix D, Page D-5 states, "The surveillance weld [Heat 305424] has two out of four data points outside the 28 °F scatter band. Hence, the surveillance data is not credible."

In RAI 2, dated March 5, 2008, the staff requested that the applicant reconcile this statement with the statement in LRA Section 4.2.2, Page 4.2-6, the second paragraph, which indicates that the "data for the Unit 1 surveillance program weld material is deemed credible" and the similar statement in WCAP-15571, Supplement 1, Section 6.1, indicating that "the data for the surveillance program weld material is deemed credible." The staff noted that the applicant must address the fact that WCAP-15571, Appendix D, shows that two of the four surveillance data points (Capsules V and Y) for Intermediate Shell Longitudinal Weld 19-714 (Heat 305424) fall outside of the 28°F ΔRT_{NDT} scatter band.

In addition to the above discrepancy, the staff noted that in LRA Section 4.2.2, Page 4.2-6, the second paragraph states that the data for the Unit 1 surveillance program weld material was "used with a σ_Δ margin of 14°F." Likewise, WCAP-15571, Supplement 1, Section 6.1 indicates that "The [surveillance program weld] data was used with a σ_Δ margin of 14°F." Therefore, the staff also requested that the applicant reconcile these statements with the 28°F value for σ_Δ presented in LRA Table 4.2-5 for Intermediate Shell Longitudinal Weld 19-714, based on RP 2.1.

In its response to RAI 2, dated April 2, 2008, the applicant acknowledged that the statements in LRA Section 4.2.2, Page 4.2.6, the second paragraph and WCAP-15571, Supplement 1, Section 6.1 are incorrect because these statements indicate (1) that the data for the surveillance program weld material (Heat No. 305424) at Unit 1 are credible and (2) that the Unit 1 surveillance weld data were used with a σ_Δ value of 14°F. The applicant stated that corrections to LRA Section 4.2.2 were made to indicate (1) that the Unit 1 surveillance weld data are non-credible and (2) a σ_Δ value of 28°F is used for the corresponding RV weld (Intermediate Shell Longitudinal Weld 19-714).

These changes to LRA Section 4.2.2 were implemented in accordance with LRA Amendment No. 5, dated April 2, 2008. The applicant provided a regulatory commitment (Regulatory Commitment 1) to incorporate these same changes in WCAP-15571, Supplement 1, Section 6.1, by September 30, 2008.

Based on its review, the staff finds the applicant's response to RAI 2 acceptable because the applicant has corrected the information in LRA Section 4.2.2 regarding the credibility of the Unit 1 surveillance weld data and the σ_Δ value for the corresponding RV weld with LRA Amendment 5. Furthermore, the applicant has committed to correcting the statements in WCAP-15571, Supplement 1, Section 6.1, regarding the credibility of the surveillance weld and the proper value for the σ_Δ term. Therefore, the staff's concern described in RAI 2 is resolved.

The staff noted that Intermediate-to-Lower Shell Circumferential Weld 11-714 (Heat 90136) and Lower Shell Longitudinal Weld 20-714 (Heat 305414) are not represented in the Unit 1 surveillance program. However, chemistry factor and RT_{PTS} values for these welds based on RP 2.1 were reported in LRA Table 4.2-5.

In RAI 3, dated March 5, 2008, the staff requested that the applicant provide additional information concerning its use of surveillance data for these welds. First, the staff requested that the applicant confirm whether the heats for these welds are represented in the surveillance programs for St. Lucie, Unit 1 (St. Lucie 1) (Heat 90136) and Fort Calhoun (Heat 305414). Second, the staff requested that the applicant indicate whether these surveillance data sets were deemed credible in accordance with RG 1.99, Revision 2, and provide references for the documents where the analyses for determining credibility (or non-credibility) may be found. Finally, the staff requested that the applicant indicate whether the chemistry factors for these welds, based on RP 2.1 (84.8 for Intermediate-to-Lower Shell Circumferential Weld 11-714 and 223.9 for Lower Shell Longitudinal Weld 20-714), were calculated by adjusting the measured ΔRT_{NDT} values by the ratio of the chemistry factor for the vessel weld to the chemistry factor for the surveillance weld, as prescribed in RG 1.99, Revision 2. If the ΔRT_{NDT} values were properly adjusted for determining these chemistry factor values, the staff requested that the applicant provide the chemistry factor ratio adjustment factors for these welds or a reference for the document where these adjustment factors may be obtained. If the ΔRT_{NDT} values were not adjusted for determining these chemistry factor values, the staff requested that the applicant modify LRA Table 4.2-5 and the Unit 1 PTLR to include chemistry factor calculations based on RP 2.1 for these welds that account for this adjustment.

In its response to RAI 3 dated April 2, 2008, the applicant confirmed that the heats for Unit 1 Intermediate-to-Lower Shell Circumferential Weld 11-714 (Heat No. 90136) and Lower Shell Longitudinal Weld 20-714 (Heat No. 305414) are represented in the surveillance programs for sister plants St. Lucie 1 and Fort Calhoun, respectively. The applicant indicated that the credibility analysis for the St. Lucie 1 surveillance data can be found in WCAP-15446, "Analysis of Capsule 284° from the Florida Power & Light Company St. Lucie Unit 1 Reactor Vessel Radiation Surveillance Program," September 2000. This analysis determined that the surveillance data for this weld heat (Heat No. 90136) is credible. The applicant indicated that the credibility analysis for the Fort Calhoun surveillance weld Heat No. 305414 can be found in WCAP-15571, Appendix D, which also provides the credibility analyses for the surveillance materials at Unit 1. This analysis determined that the surveillance data for this weld is not credible. The staff reviewed the above credibility analyses for the St. Lucie 1 and Fort Calhoun surveillance weld data and determines that the applicant accurately assessed the credibility of the surveillance data for these weld heats.

With respect to the chemistry factor ratio adjustments to the measured ΔRT_{NDT} values for this surveillance data, the applicant indicated that the chemistry factor ratio adjustment procedure was applied to account for chemistry differences between the Unit 1 RV welds and the St. Lucie 1 and Fort Calhoun surveillance welds, respectively.
The applicant indicated that the chemistry factor ratio adjustment procedures can be found in WCAP-15570, "Beaver Valley Unit 1 Heatup and Cooldown Limit Curves for Normal Operation," Revision 2, which documents the P-T limit curve calculations for Unit 1. The staff independently confirmed the validity of these adjustments and that they were correctly applied to the measured ΔRT_{NDT} values for these surveillance welds.

Based on its review, the staff finds the applicant's response to RAI 3 acceptable because the applicant has provided all of the requested information and supporting documentation regarding its analysis and implementation of surveillance weld data for St. Lucie 1 Heat No. 90136 and Fort Calhoun Heat No. 305414. Therefore, the staff's concern described in RAI 3 is resolved.

The applicant indicated in LRA Section 4.2.2 that a neutron flux management program is in place at Unit 1 for ensuring that the limiting RV beltline material, Lower Shell Plate B6903-1, meets the PTS screening requirements of 10 CFR 50.61 at the end of the current 40-year license term. The staff was unclear whether these same measures for managing neutron flux will maintain the RT_{PTS} value for the limiting material within the 10 CFR 50.61 PTS screening limits, until the year 2033 (43.87 EFPYs); the time, as documented in LRA Section 4.2.2, when the limiting material will reach the 270 °F PTS screening limit.

In RAI 5, dated March 5, 2008, the staff requested that the applicant indicate whether the limiting material is projected to reach the 270°F screening limit requirement of 10 CFR 50.61 in the year 2033 (43.87 EFPYs), under this same flux management program. If the current flux management program will not maintain the limiting material below the PTS screening limit until 2033 (43.87 EFPYs), the staff requested that the applicant discuss any additional measures required to ensure that the limiting material does not exceed the PTS screening limit until 2033 (43.87 EFPYs). Furthermore, the staff noted that LRA Section 4.2.2 states that documentation of a flux reduction program for Unit 1 will be submitted in accordance with the requirements of 10 CFR 50.61. Therefore, the staff also requested in RAI 5 that the applicant provide a formal commitment to submit the appropriate documentation of its program for maintaining the limiting RV beltline material (Plate B6903-1) at Unit 1 below the 10 CFR 50.61 PTS screening criterion through the end of the period of extended operation (54 EFPYs) and that this commitment include a schedule for submitting this documentation.

In its response to RAI 5, dated April 2, 2008, the applicant stated that the RT_{PTS} value for the limiting RV beltline material at Unit 1 (Lower Shell Plate B6903-1) will reach the 270°F screening limit (as specified in 10 CFR 50.61) at a fluence level of 4.961 x 1019 n/cm2 (E > 1.0 MeV). The applicant verified that this limiting fluence level (corresponding to an RT_{PTS} value of 270°F) will be reached for the limiting material at 43.87 EFPYs and that the limiting material RT_{PTS} value will reach 275.7°F at the end of the period of extended operation (54 EFPYs), if mitigating actions above and beyond the current neutron flux management program are not implemented. The applicant further stated that Unit 1 will reach 43.87 EFPYs in 2033, assuming a 90% capacity factor. In the second part of its response to RAI 5, the applicant stated that it has provided a formal commitment to submit the appropriate documentation of its plan for maintaining the limiting Unit 1 RV beltline material (Lower Shell Plate B6903-1) below the required 10 CFR 50.61 screening limit, through the end of the period of extended operation (54 EFPYs). This license renewal future commitment is provided in LRA Appendix A, Table A.4-1, under Item No. 24, and reads as follows:

> "Prior to exceeding the PTS screening criteria for BVPS Unit 1, FENOC [the applicant] will select a [neutron] flux reduction measure to manage PTS in accordance with the requirements of 10 CFR 50.61. A flux reduction plan will be submitted for NRC review and approval at least 1 year prior to implementation of the flux reduction measure."

The staff determines that this commitment meets the requirements of 10 CFR 50.61 for plants projected to exceed the PTS screening criteria at end-of-license under existing operating conditions. The staff further determines that 10 CFR 50.61(b)(3) explicitly permits such plants to implement future flux reduction programs that will maintain the limiting RT_{PTS} value below the applicable PTS screening limit, and that the applicant's future submittal of its flux reduction plan at least one year prior to implementation will allow sufficient time for staff review of these flux reduction measures. In addition to this commitment, the applicant revised LRA Sections 4.2.2 and A.2.2.2 (UFSAR Supplement for PTS) to indicate that the flux reduction plan will be submitted for staff review and approval at least one year prior to implementation of the flux reduction measure. These changes to LRA Sections 4.2.2 and A.2.2.2 were implemented in accordance with LRA Amendment No. 5 dated April 2, 2008.

Based on its review, the staff finds the applicant's response to RAI 5 acceptable because the applicant has provided (1) the requested information regarding the suitability of current flux management programs for maintaining the RT_{PTS} value for the limiting Unit 1 RV beltline material below the 10 CFR 50.61 screening criteria until 2033 (43.87 EFPYs), and (2) the requested formal commitment regarding a flux reduction plan to manage PTS that demonstrates compliance with the requirements of 10 CFR 50.61. Therefore, the staff's concern described in RAI 5 is resolved.

The staff noted that WCAP-16527-NP, Appendix D, "Analysis of Capsule X from First Energy Nuclear Operating Company Beaver Valley Unit 2 Reactor Vessel Radiation Surveillance Program," Revision 0 indicated that the data for the surveillance weld at Unit 2 (Heat No. 83642) were not adjusted by the ratio of the chemistry factor for the vessel weld to the chemistry factor for the surveillance weld, as prescribed in RG 1.99, Revision 2, for ART and RT_{PTS} calculations based on RP 2.1. Therefore, in RAI 7 dated March 5, 2008, the staff requested that the applicant verify whether the copper and nickel content of the surveillance weld differs from that of the RV welds at Unit 2. If a difference in chemistry exits, the staff requested that the applicant modify LRA Section 4.2.2, LRA Table 4.2-6, and the Unit 2 PTLR to account for the chemistry factor ratio adjustment.

In its response to RAI 7, dated April 2, 2008, the applicant indicated that WCAP-16527-NP, Appendix D addresses only the credibility evaluation of the surveillance materials. The Unit 2 surveillance weld data were not adjusted by the chemistry factor ratio in this report. However, weight percentage copper and nickel values for the Unit 2 RV welds are different from the weight percentage copper and nickel values for the corresponding surveillance weld (Heat No. 83642).

According to the applicant, WCAP-15677-NP, "Beaver Valley Unit 2 Heatup and Cooldown Limit Curves for Normal Operation," August 2001, documents the P-T limit curve calculations for Unit 2 which includes a RV weld to surveillance weld chemistry factor ratio calculation. WCAP-15677-NP, Tables 4-5 and 4-6, also specify best estimate values for weight percentage copper and nickel for the Unit 2 RV welds and corresponding surveillance weld. These weight percentage copper and nickel values result in chemistry factor values of 34.4 for the RV welds and 38 for the corresponding surveillance weld. The resulting RV weld to surveillance weld chemistry factor ratio is 0.905. In WCAP-15677-NP, this chemistry factor ratio was conservatively set to 1.0 for the actual RP 2.1 chemistry factor calculations. When the chemistry factor ratio is conservatively set to 1.0, the resulting RP 2.1 chemistry factor is 12.5, and the applicant used this value in LRA Section 4.2.2 for the RT_{PTS} calculations that were based on RP 2.1. The staff confirms that the actual chemistry factor value for the Unit 2 surveillance weld is 38 which results in a chemistry factor ratio of 0.905. Therefore, the applicant's RP 2.1 calculations based on a chemistry factor ratio of 1.0 for this weld are conservative.

The staff finds the applicant's response to RAI 7 acceptable because the applicant has provided sufficient explanation regarding its implementation of the chemistry factor ratio adjustment procedure for the Unit 2 RV weld embrittlement calculations. Therefore, the staff's concern described in RAI 7 is resolved.

Based on its review of the applicant's RT_{PTS} calculations and their response to RAIs 1, 2, 3, 5, and 7 dated April 2, 2008, as documented above, the staff finds that the applicant has accurately calculated the 54 EFPY RT_{PTS} values for all RV beltline materials and has correctly used applicable surveillance data for determining that the non-limiting Unit 1 RV beltline materials and all Unit 2 RV beltline materials will remain in compliance with the requirements of 10 CFR 50.61, through the end of the period of extended operation (54 EFPYs). The staff also finds that the applicant has provided an acceptable commitment regarding the future submittal of a flux reduction plan to manage PTS for the Unit 1 limiting RV beltline material that demonstrates compliance with the requirements of 10 CFR 50.61.

4.2.2.3 UFSAR Supplement

In LRA Sections A.2.2.2 (Unit 1) and A.3.2.2 (Unit 22), the applicant provided UFSAR Supplement summary descriptions for the PTS TLAA. The applicant has amended LRA Section A.2.2.2 for the Unit 1 PTS UFSAR Supplement in accordance with its April 2, 2008 RAI response. The staff reviewed the applicant's PTS UFSAR Supplement summary descriptions, as amended, for Unit 1 and Unit 2 and determines they are consistent with the PTS TLAA in LRA Section 4.2.2, as amended. The PTS UFSAR Supplements summarize the applicable 10 CFR 50.61 PTS screening requirements. The Unit 1 UFSAR Supplement summary description states that the PTS TLAA at Unit 1 will be adequately managed for the period of extended operation in accordance the requirements in 10 CFR 54.2(c)(1)(iii). The Unit 2 UFSAR Supplement summary description states that the RV beltline materials for Unit 2 will comply with the applicable requirements in 10 CFR 50.61, as projected through the end of the period of extended operation. Therefore, the staff determines that the Unit 1 and Unit 2 UFSAR Supplement summary descriptions for the PTS TLAA are acceptable.

4.2.2.4 Conclusion

Based on its review, the staff determines that (1) the PTS TLAA at Unit 1 will be managed in accordance with Commitment No. 24 to ensure compliance with 10 CFR 50.61, through the end of the period of extended operation; and (2) the RV beltline materials at Unit 2 are projected to remain in compliance with the PTS requirements in 10 CFR 50.61, through the end of the period of extended operation. The staff concludes that the applicant's (1) TLAA for PTS at Unit 1 is in compliance with 10 CFR 54.21(c)(1)(iii); (2) TLAA for PTS at Unit 2 is in compliance with 10 CFR 54.21(c)(1)(ii); and (3) safety margins established and maintained during the current operating term will be maintained during the period of extended operation, as required by 10 CFR 54.21(c)(1). The staff also concludes that the UFSAR Supplements, as amended, for Unit 1 and Unit 2 contain appropriate summary descriptions of the TLAA for PTS for the period of extended operation, as required by 10 CFR 54.21(d).

4.2.3 Charpy Upper Shelf Energy

4.2.3.1 Summary of Technical Information in the Application

In LRA Section 4.2.3, the applicant summarized the evaluation of C_vUSE for the period of extended operation. According to RG 1.99, Figure 2, without surveillance data, C_vUSE

presumably decreases as a function of fluence and copper content. Linear interpolation is permitted. With surveillance data, the decrease in C_vUSE may be obtained by plotting the reduced plant surveillance data on RG 1.99, Revision 2, Figure 2, and fitting the data with a line drawn parallel to the existing lines as the upper bound of all the data. This line should be preferred to the existing graph. The C_vUSE is predictable by use of the corresponding T/4 fluence projection, the copper content of the beltline materials, the results of the capsules tested to date by RG 1.99, Revision 2, Figure 2, or a combination of each. RV beltline materials must have an initial C_vUSE of no less than 75 ft-lb and must maintain C_vUSE of no less than 50 ft-lb throughout the life of the vessel.

4.2.3.1.1 Unit 1

In the spring of 2000, Surveillance Capsule Y was pulled for analysis documented in WCAP-15571. For license renewal, WCAP-15571, Supplement 1, documents the end-of-license-extended analysis for C_vUSE.

For Unit 1, there are material surveillance data for RV lower shell plate B6903-1 (heat C6317-1) and the intermediate shell longitudinal weld (heat 305424). The applicant plotted the measured drops in C_vUSE for each of these material heats on RG 1.99, Revision 2, Figure 2, with a horizontal line drawn parallel to the existing lines as the upper bound of all data and used RG 1.99, Revision 2, Figures 1 and 2 to determine the percent decrease in C_vUSE for the beltline and extended beltline materials. LRA Table 4.2-7 shows C_vUSE values at end-of-license-extended (54 EFPYs) for Unit 1 beltline materials. The applicant evaluated the extended beltline materials likely to receive fluence values greater than $1.0E+17$ n/cm^2 (E>1.0 MeV) and determined that none of these materials were limiting. The beltline and extended beltline material C_vUSE values maintain 50 ft-lb or greater at 54 EFPYs; therefore, disposition of the Unit 1 C_vUSE analysis complies with 10 CFR 54.21(c)(1)(ii).

4.2.3.1.2 Unit 2

In the spring of 2005, Surveillance Capsule X was pulled for analysis documented in WCAP-16527-NP. For license renewal, WCAP-16527-NP, Supplement 1, documents the end-of-license-extended analysis for C_vUSE. For Unit 2, there are material surveillance data for RV intermediate shell plate B9004-2 (heat C0544-2) and the intermediate shell longitudinal weld (heat 83642). The applicant plotted the measured drops in C_vUSE for each of these material heats on RG 1.99, Revision 2, Figure 2, with a horizontal line drawn parallel to the existing lines as the upper bound of all data and used RG 1.99, Revision 2, Figures 1 and 2 to determine the percent decrease in C_vUSE for the beltline and extended beltline materials.

LRA Table 4.2-8 shows C_vUSE values at end-of-license-extended (54 EFPYs) for the Unit 2 beltline materials. The beltline material C_vUSE values maintain 50 ft-lb or greater at 54 EFPYs. The applicant also evaluated extended beltline materials likely to receive fluence values greater than $1.0E+17$ n/ cm^2 (E>1.0 MeV). The extended beltline material C_vUSE values maintain 50 ft-lb or greater at 54 EFPYs; therefore, disposition of the Unit 2 C_vUSE analysis complies with 10 CFR 54.21(c)(1)(ii).

4.2.3.2 Staff Evaluation

The staff reviewed LRA Section 4.2.3 to verify, pursuant to 10 CFR 54.21(c)(1)(ii), that the analyses have been projected to the end of the period of extended operation.

Appendix G of 10 CFR 50 provides fracture toughness requirements for ferritic materials (low alloy steel or carbon steel) in the RCPB, including C_vUSE requirements for ensuring adequate safety margins against ductile tearing. The staff's acceptance criteria are based on (1) GDC-14, which requires that the RCPB be designed, fabricated, erected, and tested so as to have an extremely low probability of rapidly propagating fracture; (2) GDC-31, which requires that the RCPB be designed with a safety margin sufficient to assure that, under specified conditions, it will behave in a non-brittle manner and the probability of a rapidly propagating fracture is minimized; (3) 10 CFR Part 50, Appendix G, which specifies fracture toughness requirements for ferritic components of the RCPB; and (4) 10 CFR 50.60, which requires compliance with 10 CFR Part 50, Appendix G.

Appendix G of 10 CFR 50 also provides the staff's criteria for maintaining acceptable levels of C_vUSE for RV beltline materials throughout the licensed operational lives of reactor facilities. The rule requires RV beltline materials to have a minimum C_vUSE value of 75 ft-lb in the unirradiated condition, and to maintain a minimum C_vUSE value above 50 ft-lb throughout the life of the facility; unless, it can be demonstrated through analysis that lower values of C_vUSE would provide acceptable margins of safety against fracture equivalent to those required by the American Society of Mechanical Engineers (ASME) Code Section XI, Appendix G. The rule also mandates that the methods used to calculate C_vUSE values must account for the effects of neutron radiation on the C_vUSE values for the materials and must incorporate any relevant RV surveillance capsule data that are reported through implementation of a plant's RV materials surveillance program, pursuant to 10 CFR 50, Appendix H. The staff's recommended guidelines for calculating the effects of neutron radiation on the C_vUSE for the RV beltline materials are specified in RG 1.99, Revision 2. The C_vUSE value for a material at a given fluence level can be determined based on the initial (unirradiated) C_vUSE value for the material and a percentage decrease in C_vUSE that may be calculated using the procedures in RG 1.99, Revision 2. The percentage decrease in C_vUSE may be determined using RG 1.99, Revision 2, Figure 2, in accordance with RP 1.2 or from credible surveillance data pursuant to RP 2.2.

The C_vUSE for a material decreases as a function copper content and neutron fluence. Since neutron fluence changes with time, the determination of C_vUSE complies with 10 CFR 54.3(a) for being a TLAA.

In LRA Section 4.2.3, the applicant discussed the Unit 1 and Unit 2 C_vUSE calculations for the period of extended operation (54 EFPYs). The applicant noted that the Unit 1 C_vUSE calculations address surveillance data from the analysis of Surveillance Capsule Y, which was pulled from the Unit 1 RV in the spring of 2000. Likewise, the C_vUSE calculations for Unit 2 address surveillance data from the analysis of Surveillance Capsule X, which was pulled from the Unit 2 RV in the spring of 2005. The applicant stated that all of the RV beltline and extended beltline materials at Unit 1 and Unit 2 are projected to maintain C_vUSE values greater than 50 ft-lb through the end of the period of extended operation.

The applicant provided 54 EFPY C_vUSE calculations for the Unit 1 and Unit 2 RV beltline materials in LRA Tables 4.2-7 and 4.2-8. These tables included all the input data required to determine the C_vUSE values at the end of the period of extended operation, including the weight percentage copper, initial C_vUSE values, and 1/4T fluence values. The staff independently confirmed that the applicant utilized valid weight percent copper and initial C_vUSE values for the Unit 1 and Unit 2 RV beltline materials. The applicant calculated the projected 54 EFPY C_vUSE values in accordance with RG 1.99, Revision 2, RP 2.2, for all RV beltline materials represented in the RV surveillance programs at Unit 1 and Unit 2. C_vUSE calculations for all other RV beltline materials at Unit 1 and Unit 2 were based on RG 1.99, Revision 2, RP 1.2. Surveillance data from sister plants, St. Lucie 1 and Fort Calhoun, were not used in the Unit 1 C_vUSE calculations. The staff also independently confirmed that the applicant correctly calculated the 54 EFPY C_vUSE values and that all of these C_vUSE values are greater than 50-lb, as required by 10 CFR Part 50, Appendix G. In addition, the staff calculated 54 EFPY C_vUSE values based on RG 1.99, Revision 2, RP 1.2 for all of the Unit 1 and Unit 2 RV beltline materials represented in the RV surveillance programs (*i.e.*, those materials with C_vUSE values calculated using RP 2.2). This was done in order to confirm that these materials would meet the 50-ft-lb C_vUSE requirement even without application of the surveillance data, which yield non-conservative results. With the exception of Lower Shell Plate B6903-1 at Unit 1, all RV beltline materials represented in the applicant's surveillance programs maintained C_vUSE values greater than 50 ft-lb, when calculated using RP 1.2. For Unit 1, the staff found that the C_vUSE for Lower Shell Plate B6903-1 (the limiting material) when calculated using RP 2.1, is approximately 49.8 ft-lb (*i.e.* slightly less than the minimum allowable value). The applicant calculated a C_vUSE value of 51.5 ft-lb for the Unit 1 limiting plate based on RP 2.2. Therefore, in RAI 6, dated March 5, 2008, the staff requested that the applicant explain why the surveillance data were deemed credible for determining the 54 EFPY C_vUSE value for this plate, based on the criteria for surveillance data credibility from RG 1.99, Revision 2.

In its response to RAI 6, dated April 2, 2008, the applicant stated that the scatter of the C_vUSE surveillance data for Lower Shell Plate B6903-1 is small enough to permit the determination of the C_vUSE unambiguously. The applicant indicated that WCAP-15571, Supplement 1, documents the calculation of the 54 EFPY C_vUSE value for this plate, based on application of the surveillance data in accordance with RP 2.2. The applicant discussed this calculation in detail. The staff had previously confirmed the accuracy of this calculation; however, it was clear to the staff upon closer review that the C_vUSE surveillance data for this plate was sufficient to result in a reliable calculation of the 54 EFPY C_vUSE.

Based on its review, the staff concludes that the applicant has correctly determined that the surveillance data for Lower Shell Plate B6903-1 was credible for application to the C_vUSE calculation for this plate. Therefore, the staff's concern in RAI 6 is resolved.

Based on its review of the applicant's C_vUSE calculations and response to RAI 6 as documented above, the staff finds that the applicant has accurately calculated the 54 EFPY C_vUSE values for all RV beltline materials. The staff also finds the applicant has correctly used applicable surveillance data for determining that the Unit 1 and Unit 2 RV beltline materials will maintain C_vUSE values greater than 50 ft-lb, through the end of the period of extended operation (54 EFPYs), in accordance 10 CFR Part 50, Appendix G. Therefore, the staff's concern described in RAI 6 is resolved.

4.2.3.3 UFSAR Supplement

In LRA Sections A.2.2.3 (Unit 1) and A.3.2.3 (Unit 2), the applicant provided UFSAR Supplement summary descriptions for the TLAA of the C_vUSE. The staff reviewed the applicant's C_vUSE UFSAR Supplement summary descriptions for Unit 1 and Unit 2 and determines they are consistent with the TLAA for the C_vUSE in LRA Section 4.2.3. The UFSAR Supplement summary descriptions summarize the applicable C_vUSE requirements that must be met to ensure continued compliance with 10 CFR 50, Appendix G. They also state that the RV beltline materials for Unit 1 and Unit 2 will comply with the applicable requirements in 10 CFR 50, Appendix G, as projected through the end of the period of extended operation. Therefore, the staff determines that the Unit 1 and Unit 2 UFSAR Supplement summary descriptions for the TLAA on C_vUSE are acceptable.

4.2.3.4 Conclusion

Based on its review, the staff determines that the applicant projects that the RV beltline materials at Unit 1 and Unit 2 will remain in compliance with the C_vUSE requirements in 10 CFR 50, Appendix G, through the end of the period of extended operation. Therefore, the staff concludes that the applicant's TLAA for the C_vUSE is in compliance with 10 CFR 54.21(c)(1)(ii) and that the safety margins established and maintained during the current operating term will be maintained during the period of extended operation, as required by 10 CFR 54.21(c)(1). The staff also concludes that the UFSAR Supplements for Unit 1 and Unit 2 contain appropriate summary descriptions of the TLAA of the C_vUSE for the period of extended operation, as required by 10 CFR 54.21(d).

4.2.4 Pressure-Temperature Limits

4.2.4.1 Summary of Technical Information in the Application

In LRA Section 4.2.4, the applicant summarized the evaluation of P-T limits for the period of extended operation. Appendix G of 10 CFR Part 50 requires that RV boltups, hydrotests, pressure tests, normal operations, and anticipated operational occurrences be accomplished within P-T limits established by calculations that utilize the materials and fluence data from the reactor surveillance capsule analyses. P-T limits calculated for several years into the future remain valid for an established period of time, not to exceed the current operating license term.

The applicant's P-T limit curves are operating limits, conditions of the operating license, and are included in the pressure and temperature limits report, as required by TSs. They are valid up to a stated vessel fluence limit and must be revised prior to operation beyond that limit. The latest PTLR submitted to the staff for each unit was on March 31, 2005. The power uprate review evaluated the continued applicability of each unit's P-T limits.

Appendix G of 10 CFR Part 50 requires the applicant to operate within the currently licensed P-T limit curves, which must be maintained and updated as necessary for plant operation in accordance with 10 CFR Part 50. The Reactor Vessel Integrity Aging Management Program will maintain the P-T limit curves for both units for the period of extended operation; therefore,

disposition of the Unit 1 and Unit 2 P-T limit curves TLAAs is in accordance with
10 CFR 54.21(c)(1)(iii).

The applicant states that the Low-Temperature Overpressure Protection System is known as
the Overpressure Protection System (OPPS). Updates for both units reviews the OPPS
setpoints (temperature and power-operated relief valve setpoints), as required, based on the
updated P-T limit curves.

4.2.4.2 Staff Evaluation

The staff reviewed LRA Section 4.2.4 to verify, pursuant to 10 CFR 54.21(c)(1)(iii), that the
effects of aging on the intended function(s) will be adequately managed for the period of
extended operation.

Appendix G of 10 CFR Part 50 provides fracture toughness requirements for ferritic materials
(low alloy steel or carbon steel) materials in the RCPB, including requirements for calculating P-
T limits for the plant. 10 CFR Part 50, Appendix G requires that RCPB materials satisfy the
criteria in ASME Code Section XI, Appendix G, in order to ensure the structural integrity of the
RCPB during any condition of normal operation, including anticipated operational occurrences
and hydrostatic tests. Acceptance criteria for P-T limits are based on (1) GDC-14, which
requires that the RCPB be designed, fabricated, erected, and tested as to have an extremely
low probability of rapidly propagating fracture; (2) GDC-31, which requires that the RCPB be
designed with a safety margin sufficient to assure that, under specified conditions, it will behave
in a nonbrittle manner and the probability of a rapidly propagating fracture is minimized; (3)
10 CFR Part 50, Appendix G, which specifies fracture toughness requirements for ferritic
components of the RCPB; and (4) 10 CFR 50.60, which requires compliance with
10 CFR Part 50, Appendix G.

Section IV.A.2 of 10 CFR Part 50, Appendix G requires that P-T limits for operating reactors be
at least as conservative as those that would be generated using the calculation methods
specified in ASME Code Section XI, Appendix G. The rule also requires that P-T limit
calculations account for the effects of neutron radiation on the properties of the RV beltline
materials and that these calculations incorporate any relevant RV surveillance capsule data that
are required for compliance as part of the applicant's implementation of its 10 CFR Part 50,
Appendix H RV materials surveillance program. The staff's recommended guidelines for
calculating the effects of neutron radiation on the RV beltline material properties, specifically the
ART values used for calculating P-T limits, are specified in RG 1.99, Revision 2. P-T limits are
usually calculated based on the ART value for the limiting RV beltline material. The ART for a
material increases as a function of neutron fluence. Since neutron fluence changes with time,
the P-T limits meet the requirements of 10 CFR 54.3(a) for being a TLAA.

In LRA Section 4.2.4, the applicant discussed the P-T limits for Unit 1 and Unit 2 which are
documented in the Unit 1 and Unit 2 PTLRs, "Beaver Valley Power Station Unit No. 1 Pressure
and Temperature Limits Report," Revision 4, September 19, 2007, and "Beaver Valley Power
Station Unit No. 2 Pressure and Temperature Limits Report," Revision 2, June 25, 2007,
respectively. The contents of the Unit 1 and Unit 2 PTLRs are controlled in accordance with TS
requirements. The TSs for Unit 1 and Unit 2 require that the applicant operate the reactor
coolant system (RCS) within the limits specified in the PTLRs and that PTLRs be updated for
each RV fluence period or for any revision or supplement thereto. The P-T limits in the current

revisions of the Unit 1 and Unit 2 PTLRs are valid up to the RV fluence levels corresponding to operating periods explicitly specified in the reports. In accordance with TS requirements, the applicant will update the PTLRs for new fluence limits prior to operating beyond the current periods. The Unit 1 and Unit 2 TS requirements concerning RCS operation and the contents of the PTLRs ensure that the structural integrity of the RCPB will be maintained in accordance with 10 CFR Part 50, Appendix G. Additionally, the applicant specified that the Reactor Vessel Integrity Program described in LRA Section B.2.35 will maintain the Unit 1 and Unit 2 P-T limit curves to ensure compliance with 10 CFR Part 50, Appendix G, through the end of the period of extended operation (54 EFPYs).

The staff reviewed the contents of the Unit 1 and Unit 2 PTLRs and determines that, in general, they contain the data necessary to ensure compliance with TS requirements and the requirements of 10 CFR Part 50, Appendix G. However, the staff noted several anomalies in the PTLRs requiring further clarification from the applicant. Unit 1 PTLR Table 5.2-5 states that the chemistry factor for Lower Shell Plate B6903-1 is 147.2 (based on RP 1.1). This is the incorrect chemistry factor for this plate, per the February 12, 1998, NRC-Industry meeting, where the NRC recommended that the non-credible surveillance data for this *specific* plate be used along with a full σ_Δ value of 17 °F for RT_{PTS} and ART calculations. LRA Section 4.2.2 accurately reflects that the non-credible surveillance data and full σ_Δ value of 17°F were used to arrive at a 54 EFPY RT_{PTS} value of 275.7°F, based on a RP 2.1 chemistry factor value of 149.2. Furthermore, PTLR Table 5.2-7 provides ART calculations for this limiting plate that are based on the correct chemistry factor value of 149.2 and states that these calculations are based on the non-credible plate surveillance data and full σ_Δ value of 17°F. The staff noted that the application of surveillance data and the selection of chemistry factors for calculation of RT_{PTS} and ART values in the Unit 1 PTLR should be consistent with the LRA. Therefore, in RAI 4 dated March 5, 2008, the staff requested that the applicant explain why Unit 1 PTLR Table 5.2-5 showed a chemistry factor value of 147.2 for Lower Shell Plate B6903-1 instead of the correct chemistry factor value of 149.2.

In its response to RAI 4, dated April 2, 2008, the applicant stated that Unit 1 PTLR Table 5.2-5 should not show a chemistry factor value 149.2 for Lower Shell Plate B6903-1 because this table only applies to chemistry factor calculations based on RG 1.99, Revision 2, RP 1.1. A chemistry factor value of 147.2, based on RP 1.1, is the correct chemistry factor value for this table. The applicant indicated that the Unit 1 PTLR Table 5.2-4 shows chemistry factor calculations based on the use of surveillance data. The staff reviewed the Unit 1 PTLR Table 5.2-4 and finds that the chemistry factor for Lower Shell Plate B6903-1 was correctly calculated at 149.2, based on the application of the non-credible surveillance data with a full σ_Δ value of 17°F. Furthermore, the staff noted that the titles for these tables adequately reflect the applicant's intent to calculate chemistry factors in accordance with both RPs found in RG 1.99, Revision 2, for RV beltline materials represented in the surveillance program. This strategy, whereby the applicant calculates chemistry factors based on RPs 1.1 and 2.1 and, selects the more conservative factor, is consistent with the LRA chemistry factor calculations for the PTS TLAA discussed in SER Section 4.2.2.

For Unit 2, the staff noted that LRA Section 4.2.2, LRA Table 4.2-6, and WCAP-16527-NP, Supplement 1 all incorporate data from the evaluation of Surveillance Capsules U, V, W, and X. However, the staff noted that the Unit 2 PTLR only incorporates data from the evaluation of Surveillance Capsules U, V and W. The application of surveillance data and the selection of chemistry factors for calculation of RT_{PTS} and ART values in the Unit 2 PTLR should be consistent with the LRA. As the Unit 2 PTLR forms part of the basis for the LRA, the staff requested in RAI 8, dated March 5, 2008, that the applicant update the Unit 2 PTLR to incorporate the results from the evaluation Surveillance Capsule X.

In its response to RAI 8, dated April 2, 2008, the applicant stated that the latest Unit 2 PTLR (Revision 2) was associated with implementation of the applicant's Improved TS Conversion License Amendment for Unit 2. Therefore, the Unit 2 PTLR only incorporated data from the evaluation of Surveillance Capsules U, V, and W. The applicant addressed the discrepancy between the LRA and the current revision of the Unit 2 PTLR by adding a regulatory commitment to submit to the staff by September 30, 2008, an updated PTLR that incorporates the results from the analysis of Surveillance Capsule X. This commitment was provided as Commitment No. 2 in Enclosure 1 to the applicant's April 2, 2008 RAI response. The staff reviewed this commitment and determines that the applicant has ensured that the Unit 2 PTLR will be updated in a timely manner to incorporate the results from the analysis of Surveillance Capsule X. Therefore, the staff's concern described in RAI 8 is resolved.

Based on its review of the applicant's P-T limits TLAA in LRA Section 4.2.4, Unit 1 and Unit 2 PTLRs, and the applicant's responses to RAIs 4 and 8, as documented above, the staff finds that the applicant has adequately demonstrated that the P-T limits at Unit 1 and Unit 2 will be managed to ensure compliance with 10 CFR Part 50, Appendix G through the end of the period of extended operation. Therefore, the staff's concerns described in RAIs 4 and 8 are resolved.

4.2.4.3 UFSAR Supplement

In LRA Sections A.2.2.4 (Unit 1) and A.3.2.4 (Unit 2), the applicant provided UFSAR Supplement summary descriptions for the P-T limits TLAA. The staff reviewed the applicant's P-T limits UFSAR Supplement summary descriptions for Unit 1 and Unit 2 and determines they are consistent with the TLAA for the P-T limits in LRA Section 4.2.3. The UFSAR Supplement summary descriptions summarize the applicable fracture toughness requirements that must be met to ensure continued compliance with 10 CFR Part 50, Appendix G. They also state that the P-T limits for Unit 1 and Unit 2 will be managed through the implementation of the Reactor Vessel Integrity Aging Management Program to ensure compliance with 10 CFR Part 50, Appendix G, through the end of the period of extended operation. Therefore, the staff determines that the Unit 1 and Unit 2 UFSAR Supplement summary descriptions for the P-T limits TLAA are acceptable.

4.2.4.4 Conclusion

The staff reviewed the applicant's TLAA for the P-T limits, as summarized in LRA Section 4.2.4, including its RAI responses dated April 2, 2008, and determines that the P-T limits at Unit 1 and Unit 2 will be managed through the applicant's implementation of the Reactor Vessel Integrity Aging Management Program to ensure compliance with the requirements in 10 CFR Part 50, Appendix G, through the end of the period of extended operation. Therefore, the staff concludes that the applicant's TLAA for the P-T limits is in compliance with the acceptance criterion for TLAAs pursuant to 10 CFR 54.21(c)(1)(iii) and that the safety margins established and maintained during the current operating term will be maintained during the period of extended operation, as required by 10 CFR 54.21(c)(1). The staff further concludes that the UFSAR

Supplements for Unit 1 and Unit 2 contain appropriate summary descriptions of the P-T limits TLAA for the period of extended operation, as required by 10 CFR 54.21(d).

4.3 Metal Fatigue

Unit 1 Class 1 components evaluated for fatigue include:

- Reactor vessel
- Control rod drive mechanisms
- Reactor vessel internals
- Pressurizer
- Replacement steam generators
- Reactor coolant pumps
- Loop stop valves

The applicant stated in the LRA that the design and analysis of the Unit 1 main coolant loop piping, including the pressurizer surge line, initially complied with American National Standards Institute (ANSI) B31.1. The reanalysis of the pressurizer surge line complied with ASME Code Section III to account for stratification issues in accordance with the guidance in NRC Bulletin 88-11. No other Unit 1 piping systems were designed and analyzed pursuant to ASME Code Section III.

Unit 2 Class 1 components evaluated for fatigue include:

- Reactor vessel

- Control rod drive mechanisms

- Reactor vessel internals

- Pressurizer

- Steam generators

- Reactor coolant pumps

- Loop stop valves

- Piping (main coolant loop piping, pressurizer surge line, pressurizer safety and relief valve piping, and Class 1 portions of various systems (e.g., residual heat removal (RHR), chemical and volume control, and safety injection) integral with the RCS

The Unit 2 reactor head vent and RV level instrumentation system piping also comply with ASME Code Section III Class 1, but are exempt from full fatigue analysis as they are 1-inch or less diameter.

Non-Class 1 component types within the scope of license renewal evaluated for fatigue include:

- Piping
- Tubing
- Fittings
- Tanks
- Vessels

- Heat exchangers
- Valve bodies and bonnets
- Pump casings
- Miscellaneous process components

10 CFR 54.21(c) requires an evaluation of TLAAs to demonstrate that the analyses remain valid for the period of extended operation, the analyses have been projected to the end of the period of extended operation, or the effects of aging on the intended function(s) will be adequately managed for the period of extended operation.

LRA Section 4.3 includes the following information:

- Section 4.3.1 addresses Class 1 fatigue TLAAs.

- Section 4.3.2 addresses Non-Class 1 fatigue TLAAs.

- Section 4.3.3 addresses both the fatigue TLAAs responsive to NRC Bulletins 88-08 and 88-11 and the effects of the primary coolant environment on fatigue life.

- Section 4.3.4 addresses the transients for calculation of fatigue usage factors for the ASME Code Class 1 components. For this set of cyclic design transients, the applicant compiled the cycles accrued to October 2003 and projected the cycles expected at the end of 60 years of operation to keep the results below the number of design-allowable cycles.

4.3.1 Class 1 Fatigue

The applicant stated that the design of Unit 1 and Unit 2 Class 1 components incorporates the ASME Code Section III requirement of a discrete analysis of the thermal and dynamic stress cycles on components that make up the RCPB. The fatigue analyses rely on the definition of "design-basis transients" that envelope the expected cyclic service and the calculation of a cumulative usage factor (CUF). In accordance with ASME Code Section III, Subsection NB, the CUF must not exceed 1.0. The applicant also stated that the required analysis for Unit 1 and Unit 2 incorporated a set of design-basis transients based on the original 40-year operating life of the plant. The ASME Code Section III, Class 1 fatigue evaluations in the specific piping and component analyses are TLAAs based on a number of design cycles assumed for the life of each plant.

In UFSAR Tables 4.1-10 and 3.9N-1, the applicant showed the original design-basis transients, including RCS design cycles for Unit 1 and Unit 2 and replicated those findings in LRA Table 4.3-2, which also lists operational cycles anticipated to occur during 60 years of plant life. The applicant further stated that it reviewed the design cycles against 60-year projected operational cycles and determined that the design cycles are bounding for the period of extended operation, except in specific cases described in the following three subsections. The applicant concluded that since it used the 60-year projected operational cycles to determine that the design fatigue analyses remain valid for 60 years, the Metal Fatigue of Reactor Coolant Pressure Boundary Program must continue to validate the assumptions for these analyses. Therefore, disposition of Class 1 components and piping fatigue TLAAs, except in specific cases described in the following sections, is in accordance with 10 CFR 54.21(c)(1)(i) and 10 CFR 54.21(c)(1)(iii).

4.3.1.1 Unit 2 RHR Piping and Unit 2 Charging Line

4.3.1.1.1 Summary of Technical Information in the Application

In LRA Section 4.3.1.1, the applicant summarized its evaluation of Unit 2 RHR piping and the Unit 2 charging line for the period of extended operation. The applicant stated that projected Unit 2 RHR piping and the Unit 2 charging line cycles of operation will exceed their design cycles during the period of extended operation. The Metal Fatigue of Reactor Coolant Pressure Boundary Program will monitor the transient cycles for the Unit 2 RHR piping and the Unit 2 charging line. The applicant also stated that the program will take corrective actions as required to ensure that the design bases of the these components are not exceeded for the period of extended operation and concluded that the disposition of the Unit 2 RHR piping and the Unit 2 charging line fatigue TLAAs complies with 10 CFR 54.21(c)(1)(iii).

4.3.1.1.2 Staff Evaluation

The staff reviewed LRA Section 4.3.1.1 to verify, pursuant to 10 CFR 54.21(c)(1)(iii), that the effects of aging on the intended function(s) will be adequately managed for the period of extended operation.

In LRA Table 4.3-2, the applicant provided the cycle counts as of October 15, 2003, and the estimated cycle counts at 60 years of plant operation for design transients. The applicant indicated that the design cycles are bounding for the period of extended operation, except in certain cases. The staff noted the statement in LRA Section 4.3.1.1, "Unit 2 RHR piping and Unit 2 charging line cycles of operation are projected to exceed their respective design cycles during the period of extended operation." In RAI B.2.27-4, dated May 28, 2008, the staff requested that the applicant justify the discrepancy between the text in the LRA onsite basis documents and LRA Table 4.3-1, Annotation (a). In addition, in RAI B.2.27-7, also dated May 28, 2008, that staff requested that the applicant (1) specify the major components affected by the critical and supplemental transients and, confirm that the fatigue analysis on these components has been updated to include these transients; (2) justify the consistency between those supplemental transients and design transients noted in the design specification and; (3) explain the method for monitoring these transients and indicate whether the number of design cycles for the supplemental transients will remain valid for the period of extended operation.

In response to RAI B.2.27-4, dated July 11, 2008, the applicant stated that for the location with the Annotation (a), RHR System Piping, the transient of concern is "Placing RHR in Service" and occurs at approximately 350 °F, during plant shutdown procedures. As documented in WCAP-16173-P, the applicant stated that Westinghouse initially counted this transient assuming an occurrence each time the plant transitioned from Mode 3 (Hot Standby) to Mode 4 (Hot Shutdown). The staff verified in the applicant's UFSAR and TSs that the RHR was placed into service during the transition between Mode 3 and Mode 4. The applicant noted that this method of counting transients is dependent on an accurate account of the plant modes and the transition between Mode 3 and Mode 4. To obtain an accurate count of the plant mode history, the applicant evaluated data obtained from Power Ascension Testing, through October 15, 2003. The applicant's recount analysis resulted in 31 events compared to the Westinghouse count of 85 events. The staff compared the results of the applicant's recount with LRA Table 4.3-2 and noted that Unit 2 had 30 plant cooldown cycles.

The staff further noted that the new recount performed by the applicant was reasonable because the transient "Placing RHR in Service" would have occurred at least every time the plant experienced the transient "Plant Cooldown" (*i.e.*, when the plant transitioned from Mode 3 to Mode 4). On this basis, the staff finds the applicant's response acceptable.

In response to RAI B.2.27-7, dated July 11, 2008, the applicant indicated that all the supplemental transients listed in the LRA are applicable to both Unit 1 and Unit 2. The applicant identified those components affected by each of the transients (pressurizer insurge/outsurge, selected chemical and volume control system (CVCS), auxiliary feedwater (AFW) injection and RHR activation). The staff noted that the applicant specified the applicable analyses for the components and incorporated the corresponding transients affecting them. Therefore, with the exception of the ASME Class 1 portion of the Unit 2 charging piping, no revision is required. The applicant committed (Regulatory Commitment No. 1) to perform reanalyses for the ASME Class 1 portion of the Unit 2 charging piping and to incorporate the revised design cycles of the selected CVCS transients. The applicant stated the AFW injection transient was incorporated into the original analysis for the Unit 2 reactor coolant pumps, pressurizer and loop stop valves. However, Westinghouse did not identify this transient in the nuclear steam supply system (NSSS) transients; therefore, it was not a part of the original design basis. The applicant specifically added this transient for the steam generators as part of the design basis for the extended power uprate. The staff noted that the RHR activation for Unit 2 was part of the original design basis and considered a supplemental transient, because the applicant expected that the cycles would exceed the design cycles. However, based on its response to RAI B.2.27-4, the applicant no longer expects these cycles to exceed the design cycles.

The staff noted in the applicant's response to RAI B.2.27-7, that the applicant is capable of monitoring the pressurizer insurge/outsurge, selected CVCS and AFW injection transients with its plant computer data archiving system. The staff noted that with the use of the plant computer, the applicant can identify the pressurizer insurge/outsurge transient via the surge line thermocouple, which detects a delta-temperature, and allocate it into a pre-existing band of delta-temperatures. The applicant explained that the plant computer identifies selected CVCS transients by noting the valve positions and the AFW injection transient by noting the operation and system flow rates of the AFW pumps during Plant Mode 1, 2 and 3. As discussed in the staff's evaluation of RAI B.2.27-4, RHR activation can be identified when the plant transitions between Mode 3 and Mode 4.

Based on its review of the applicant's responses to RAIs B.2.27-4 and B.2.27-7, the staff finds that the applicant has provided sufficient detail pertaining to the supplemental transients, the components affected by these transients and the method for monitoring and identifying these transients, through the period of extended operation. By letter dated October 2, 2008, the applicant completed the re-analysis and provided the results and methodology which demonstrated that the CUF, including environmental factors for the BVPS charging piping will remain below the code allowable limit of 1.0. The staff noted this revised analysis incorporated new and revised thermal transients reflecting the operating experience at BVPS Unit 2. Therefore, the staff's concerns described in RAIs B.2.27-4 and B.2.27-7 are resolved.

4.3.1.1.3 UFSAR Supplement

The applicant provided a UFSAR supplement summary description of its TLAA evaluation of Unit 2 RHR piping and Unit 2 charging line in LRA Section A.3.3.1.1. Based on its review of the UFSAR supplement, the staff concludes that the summary description of the applicant's actions to address Unit 2 RHR piping and Unit 2 charging line is adequate.

4.3.1.1.4 Conclusion

Based on its review, as discussed above, the staff concludes that the applicant has demonstrated, pursuant to 10 CFR 54.21(c)(1)(iii), that, for Unit 2 RHR piping and Unit 2 charging line, the effects of aging on the intended function(s) will be adequately managed for the period of extended operation. The staff also concludes that the UFSAR supplement contains an appropriate summary description of the TLAA evaluation, as required by 10 CFR 54.21(d).

4.3.1.2 Unit 2 Steam Generator Manway Bolts and Tubes

4.3.1.2.1 Summary of Technical Information in the Application

LRA Section 4.3.1.2 summarizes the evaluation of Unit 2 steam generator manway bolts and tubes for the period of extended operation. The applicant could not demonstrate the validity of the original design fatigue calculations through the period of extended operation for the following Unit 2 steam generator subcomponents:

- Steam generator secondary manway bolts
- Steam generator tubes (U-bend fatigue)

The applicant stated that the Unit 2 steam generator secondary manway bolt and the steam generator tube fatigue analyses are based on a 40-year life (current operating license expires in 2027). The extended power uprate temperature average coastdown analysis for the secondary manway bolts assumed replacement of the Unit 2 steam generators by the year 2027. The uprate analyses for the U-bends assumed replacement of the Unit 2 steam generators by the year 2027. The applicant further stated that the Steam Generator Tube Integrity Program will reanalyze, repair, or replace the affected components so their design bases are not exceeded for the period of extended operation; therefore, disposition of the Unit 2 steam generator secondary manway bolts and the Unit 2 steam generator tubes fatigue TLAAs complies with 10 CFR 54.21(c)(1)(iii).

4.3.1.2.2 Staff Evaluation

The staff reviewed LRA Section 4.3.1.2 to verify, pursuant to 10 CFR 54.21(c)(1)(iii), that the effects of aging on the intended function(s) will be adequately managed for the period of extended operation.

In LRA Section 4.3.1.2, the applicant indicated that it will perform a reanalysis, repair, or replacement of the affected Unit 2 steam generator manway bolts and tubes as part of an aging management program (AMP). The staff noted that the applicant also made a commitment (Commitment No. 26) in LRA Table A.5-1. However, the staff noted that the AMP description

4-28

provided in LRA Section B.2.27 does not indicate the reanalysis, repair, or replacement of the above mentioned components. In RAI 4.3.2, dated May 28, 2008, the staff requested that the applicant explain the discrepancy between LRA Section 4.3.1.2 and LRA Section B.2.27.

In its response to RAI 4.3.2, dated July 11, 2008, the applicant amended LRA Section 4.3.1.2 to indicate that the Metal Fatigue of Reactor Coolant Pressure Boundary Program will be enhanced to include reanalysis, repair or replacement of the Unit 2 steam generator manway bolts and tubes. The applicant also appropriately amended LRA Section B.2.27 and LRA Table A.5-1 to include this enhancement (Commitment No. 26). The staff further noted that as part of the extended power uprate, the applicant assumed that the steam generators would be replaced prior to year 2027, and as a result, were not reanalyzed for the period of extended operation. As such, these components will be monitored under the Metal Fatigue of Reactor Coolant Pressure Boundary Program and corrective actions, which include reanalysis, repair or replacement, will be taken in order to ensure that the design basis of these components are not exceeded during the period of extended operation.

Based on its review, the staff finds the applicant's response to RAI 4.3-2 acceptable because the applicant amended the LRA to address the discrepancy described above and has committed (Commitment No. 26) to reanalyze, repair or replace those components specified above as part of the Metal Fatigue of Reactor Coolant Pressure Boundary Program so that the appropriate corrective actions will be taken to ensure that the effects of aging on the intended functions of these components will be adequately managed for the period of extended operation. Therefore, the staff's concern described in RAI 4.3.2 is resolved.

4.3.1.2.3 UFSAR Supplement

The applicant provided a UFSAR supplement summary description of its TLAA evaluation of Unit 2 steam generator manway bolts and tubes in LRA Section A.3.3.1.2. Based on its review of the UFSAR supplement, the staff concludes that the summary description of the applicant's actions to address Unit 2 steam generator manway bolts and tubes is adequate.

4.3.1.2.4 Conclusion

Based on its review, as discussed above, the staff concludes that the applicant has demonstrated, pursuant to 10 CFR 54.21(c)(1)(iii), that, for Unit 2 steam generator manway bolts and tubes, the effects of aging on the intended function(s) will be adequately managed for the period of extended operation.

The staff also concludes that the UFSAR supplement contains an appropriate summary description of the TLAA evaluation, as required by 10 CFR 54.21(d).

4.3.1.3 Unit 1 and Unit 2 Pressurizers

4.3.1.3.1 Summary of Technical Information in the Application

In LRA Section 4.3.1.3, the applicant summarized its evaluation of Unit 1 and Unit 2 pressurizers for the period of extended operation. The applicant stated that a revision to the analysis of the Unit 1 pressurizer, lower shell, and related components in 1999 addressed improvements to the insurge/outsurge transients found by the Westinghouse Owners Group

(WOG). The applicant further stated that it had revised plant operating procedures to follow the guidance of the WOG, to minimize the impact of potential insurges. Prior to the 1999 reanalysis, Unit 1 experienced several pressurizer spray transients that challenged the analytical and TS limit of 320 °F difference between the spray line and the pressurizer steam space temperatures. The applicant incorporated revised transients in its analysis for initial spray flow and in 2005, decided to revise the operating procedures further to optimize the plant shutdown and startup processes.

The applicant stated that the optimized procedures meet all recommendations of the WOG and virtually eliminate the potential for insurges. Next, the applicant utilized the Extended Power Uprate Project to evaluate the revised uprate transients against its previous analysis. The Unit 1 pressurizer CUFs are less than 1.0.

The applicant also stated that in 2000, its revised analysis of the Unit 2 pressurizer, lower shell, and related components addressed the insurge/outsurge transients found by the WOG. Revised plant operating procedures followed the guidance of the WOG to minimize the impact of potential insurges. In 2002, the applicant decided to revise the operating procedures further to optimize the Unit 2 shutdown and startup processes. The applicant stated that its optimized procedures met all recommendations of the WOG and virtually eliminated the potential for insurges. The applicant then utilized the Extended Power Uprate Project to evaluate the revised uprate transients against the previous analysis. Because some operating parameters had changed, the applicant revised its analysis of the Unit 2 pressurizer, lower shell, and related components. In addition, the Pressurizer Weld Overlay Project had the potential to impact the pressurizer spray nozzle, the safety valve nozzles, the pressure-operated relief valve nozzle, and the surge line nozzle, during the Unit 2 refueling outage (RFO) 12 (October - November 2006). The applicant submitted a supplement to the subject analysis to address the weld overlay for the surge nozzle. The Unit 2 pressurizer CUFs are less than 1.0.

The applicant further stated that it had determined that the design fatigue analyses for the pressurizers remain valid for 60-years, using the 60-year projected operational cycles; thus, the Metal Fatigue of Reactor Coolant Pressure Boundary Program must continue to validate the assumptions in the evaluations. In addition, the pressurizer insurge cycle assumptions in the pressurizer analyses require validation for the period of extended operation. The Metal Fatigue of Reactor Coolant Pressure Boundary Program treats the pressurizer insurge as a supplemental transient requiring monitoring; therefore, disposition of the pressurizer fatigue TLAAs is in accordance with 10 CFR 54.21(c)(1)(iii).

4.3.1.3.2 Staff Evaluation

The staff reviewed LRA Section 4.3.1.3 to verify, pursuant to 10 CFR 54.21(c)(1)(iii), that the effects of aging on the intended function(s) will be adequately managed for the period of extended operation.

In LRA Section 4.3.1.3, the applicant stated that it had evaluated the pressurizer components, considering the revised extended power uprate transients, against the pervious analysis. The applicant calculated the revised CUFs associated with the pressurizer and found that they were less than the design allowable limit of 1.0. The applicant also indicated that it had used the 60-year projected operational cycles to calculate whether the design fatigue analyses for the pressurizer remains valid during the period of extended operation. In RAI B.2.27-7, dated May 28, 2008, the staff requested that the applicant (1) provide the major components affected by these transients and an update of the related fatigue analysis; (2) justify the consistency between supplemental transients and design transients; and (3) explain the method used for monitoring these transients and whether the design cycles for the supplemental transients will remain valid for the period of extended operation.

In its response to RAI B.2.27-7, dated July 11, 2008, the applicant clarified that all the supplemental transients listed in the LRA are applicable to both Unit 1 and Unit 2. The applicant continued in its response by listing the components affected by each of the transients; namely, pressurizer insurge/outsurge, selected CVCS, AFW injection and RHR activation transients. The staff noted that the applicant's analyses has incorporated the corresponding transients affecting these components and do not require revision, with the exception of the ASME Class 1 portion of the Unit 2 charging piping. The analyses for the ASME Class 1 portion of the Unit 2 charging piping is part of the applicant's commitment (Regulatory Commitment No. 1) to perform a reanalysis, incorporating the revised design cycles of the selected CVCS transients. The applicant stated that the AFW injection transient was incorporated into the original analysis for the Unit 2 reactor coolant pumps, pressurizer and loop stop valves; however, Westinghouse did not identify this transient in the NSSS transients. Therefore, the AFW injection transient was not a part of the original design basis. The applicant added this transient for the steam generators as part of the design basis for the extended power uprate. The staff noted that the RHR activation for Unit 2 was part of the original design basis, and was considered a supplemental transient because the cycles were expected to exceed the design cycles. However, based on its response to RAI B.2.27-4, the applicant no longer expects these cycles to exceed the design cycles.

The staff noted that the applicant is capable of monitoring the pressurizer insurge/outsurge, selected CVCS and AFW injection transient with its plant computer data archiving system. The staff noted that with the plant computer, the applicant can identify the pressurizer insurge/outsurge transient via the surge line thermocouple, which detects a delta-temperature, and allocate it into a pre-existing band of delta-temperatures. The applicant explained that the plant computer identifies selected CVCS transients by noting the valve positions and the AFW injection transient by noting the operation and system flow rates of the AFW pumps during Plant Modes 1, 2 and 3. As discussed in the staff's evaluation of RAI B.2.27-4, RHR activation can be identified when the plant transitions between Mode 3 and Mode 4.

Based on its review of the applicant's response to RAI B.2.27-7, the staff finds that the applicant has provided sufficient detail pertaining to the supplemental transients, the components affected by these transients and the method for monitoring and identifying these transients, through the period of extended operation. The staff further finds that the applicant has committed (Regulatory Commitment No. 1) to reanalyze the Unit 2 charging piping to incorporate the revised design cycles and has demonstrated that it is capable of identifying and monitoring critical and supplemental transients, and their associated aging effects, through the period of extended operation. Therefore, the staff's concerns described in RAI B.2.27-7 are resolved.

The staff noted that LRA Section 4.3.1.3 describes the Pressurizer Weld Overlay Project for Unit 2 only. In RAI 4.3-4, dated May 28, 2008, the staff requested that the applicant confirm whether the Pressurizer Weld Overlay Project will also be performed for Unit 1 and explain the impact of the weld overlay on the fatigue usage for Unit 1 and Unit 2 for the period of extended operation.

In its response to RAI 4.3-4, dated July 11, 2008, the applicant stated that it had completed the planned structural weld overlay for Unit 1 during RFO 18 (Fall 2007). The staff noted that the scope of work for the Unit 1 Pressurizer Weld Overlay Project was completed after the LRA was submitted. The applicant further explained that the scope of work included the pressurizer spray nozzle, relief nozzle and three safety nozzles, but did not include the pressurizer surge line. The staff confirmed in its Safety Evaluation, "Beaver Valley Power Station, Unit No. 1 – Relief Request No. BV1-PZR-01 Regarding Weld Overlay Repairs on Pressurizer Nozzle Welds (TAC No. MD4828)", dated September 17, 2007, that the pressurizer surge line was not within the scope of the project. The applicant continued to describe that for both Unit 1 and Unit 2 pressurizer nozzles, a fatigue crack growth analyses was performed using the methodology pursuant to ASME Code Section XI. The applicant determined that the impact of the structural weld overlay material on the primary stress qualifications, which include deadweight and dynamic loading, were insignificant. The applicant further stated that the pressurizer nozzles continue to meet the applicable ASME Code Section III requirements.

Based on its review, the staff finds that the applicant has adequately determined the effect of the structural weld overlay material on fatigue usage and has confirmed that the pressurizer nozzles meet the applicable requirements of ASME Code Section III. The staff concludes that the applicant's pressurizer fatigue TLAA will be part of the Metal Fatigue of Reactor Coolant Pressure Boundary Program for Unit 1 and Unit 2; and, in accordance with 10 CFR 54.21(c)(1)(iii), the effects of aging on the intended function(s) will be adequately managed for the period of extended operation. Therefore, the staff's concern described in RAI 4.3-4 is resolved.

4.3.1.3.3 UFSAR Supplement

The applicant provided a UFSAR supplement summary description of its TLAA evaluation of Unit 1 and Unit 2 pressurizers in LRA Sections A.2.3.1.1 and A.3.3.1.3. Based on its review of the UFSAR supplement, the staff concludes that the summary description of the applicant's actions to address Unit 1 and Unit 2 pressurizers is adequate.

4.3.1.3.4 Conclusion

Based on its review, as discussed above, the staff concludes that the applicant has demonstrated, pursuant to 10 CFR 54.21(c)(1)(iii), that, for Unit 1 and Unit 2 pressurizers, the effects of aging on the intended function(s) will be adequately managed for the period of extended operation. The staff also concludes that the UFSAR supplement contains an appropriate summary description of the TLAA evaluation, as required by 10 CFR 54.21(d).

4.3.2 Non-Class 1 Fatigue

Non-Class 1 component types evaluated for fatigue include pipe, tubing, fittings, tanks, vessels, heat exchangers, valve bodies and bonnets, pump casings, turbine casings, and miscellaneous process components.

4.3.2.1 Piping and In-Line Components

4.3.2.1.1 Summary of Technical Information in the Application

In LRA Section 4.3.2.1, the applicant summarized its evaluation of piping and in-line components for the period of extended operation. The applicant stated that non-Class 1 piping and in-line components (*e.g.*, fittings and valves) within the scope of license renewal comply with ANSI B31.1 or ASME Code Section III, Subsections NC and ND (*i.e.*, Class 2 or 3). These codes require the application of stress range reduction factors against the allowable stress range when evaluating cyclic secondary stresses (*i.e.*, stresses due to thermal expansion and anchor movements). Components with fewer than 7,000 cycles are limited to the calculated allowable stress range without reduction. Components likely to exceed 7,000 cycles have allowable stress ranges reduced by application of the stress range reduction factor.

The applicant also stated that for non-Class 1 components subject to cracking due to fatigue, it had reviewed system operating characteristics to determine the approximate frequency of any significant thermal cycling. If the number of equivalent full-temperature cycles is below the limit for the original design (usually 7000 cycles), the component is suitable for extended operation. If the number of equivalent full-temperature cycles exceeds the limit, evaluation of the individual stress calculations is required.

The applicant further stated that it had evaluated the validity of this assumption for 60 years of plant operation. Except for the Unit 2 emergency diesel generator (EDG) air start system, the results of this evaluation indicated that the thermal cycle assumption is valid and bounding for 60 years of operation; therefore, the non-Class 1 piping fatigue TLAAs, except the Unit 2 EDG air start subsystem fatigue TLAA, remain valid for the period of extended operation, in accordance with 10 CFR 54.21(c)(1)(i).

The Unit 2 EDG air start system has components that may be potentially subject to fatigue. The applicant will use the Metal Fatigue of Reactor Coolant Pressure Boundary Program to assess whether the full-temperature cycles limit will be exceeded in 60 years of operation. The applicant stated that with corrective actions as appropriate (including reanalysis, repair, or replacement), the full-temperature cycles of the Unit 2 EDG air start system will not be exceeded during the period of extended operation; therefore, disposition of the Unit 2 EDG air start system fatigue TLAA is in accordance with 10 CFR 54.21(c)(1)(iii).

4.3.2.1.2 Staff Evaluation

The staff reviewed LRA Section 4.3.2.1 to verify, pursuant to 10 CFR 54.21(c)(1)(i), that the analyses have been projected to the end of the period of extended operation and, pursuant to

10 CFR 54.21(c)(1)(iii), that the effects of aging on the intended function(s) will be adequately managed for the period of extended operation.

In LRA Section 4.3.2.1, the applicant discussed its evaluation of non-Class 1 components and indicated that the number of design cycles will remain bounding for the period of extended operation. Therefore, the fatigue analyses remain valid in accordance with the requirements of 10 CFR 54.21(c)(1)(i), with the exception of the Unit 2 EDG air start system. In addition, the applicant stated that for the Unit 2 EDG air start system, "BVPS will perform an assessment to

determine whether the full-temperature cycles limit will be exceeded for 60 years of operation."
In RAI 4.3-12, dated May 28, 2008, the staff requested that the applicant provide the information on the estimated temperature cycles expected for 60 years of operation and explain how these temperature cycles will be monitored during the period of extended operation.

In its response to RAI 4.3-12, dated July 11, 2008, the applicant amended LRA Section 4.3.2.1 to describe the EDG air start system as a stand-by system that operates only when the air start tank requires a top-off or refill, after it has been discharged. The staff noted that the piping for this system is subjected to heat only during the air compression cycle. The applicant has revised its design analysis to include a new load case which is representative of the observed temperatures during air compression. The applicant's evaluation verified that the stress levels from the new thermal load case are below the endurance limit for the piping material. The staff noted that the applicant has amended LRA Section 4.3.2.1 such that the TLAAs are dispositioned pursuant to 10 CFR 54.21(c)(1)(i) and (ii), only. The staff also noted that the applicant had removed Commitment No. 27, since the EDG Air Start System is now dispositioned in accordance with 10 CFR 54.21(c)(1)(ii).

Based on its review, the staff finds the applicant's response to RAI 4.3-12 acceptable because (1) the applicant has performed an evaluation that incorporates the stress levels associated with piping heating caused during the compressing of the air; (2) the results of this evaluation show that the stress levels are below the endurance limit of the piping material and thus, this TLAA is dispositioned in accordance with 10 CFR 54.21(c)(1)(ii) and the analyses projected to the end of the period of extended operation. Therefore, the staff's concern described in RAI 4.3-12 is resolved.

4.3.2.1.3 UFSAR Supplement

The applicant provided a UFSAR supplement summary description of its TLAA evaluation of piping and in-line components in LRA Sections A.2.3.2.1 and A.3.3.2.1. Based on its review of the UFSAR supplement, the staff concludes that the summary description of the applicant's actions to address piping and in-line components is adequate.

4.3.2.1.4 Conclusion

Based on its review, as discussed above, the staff concludes that the applicant has demonstrated, pursuant to 10 CFR 54.21(c)(1)(i) that the analyses remain valid for the period of extended operation and 10 CFR 54.21(c)(1)(ii), that the analyses have been projected to the end of the period of extended operation. The staff also concludes that the UFSAR supplement contains an appropriate summary description of the TLAA evaluation, as required by 10 CFR 54.21(d).

4.3.2.2 *Pressure Vessels, Heat Exchangers, Storage Tanks, Pumps, and Turbine Casings*

4.3.2.2.1 Summary of Technical Information in the Application

In LRA Section 4.3.2.2, the applicant summarized its evaluation of pressure vessels, heat exchangers, storage tanks, pumps, and turbine casings for the period of extended operation. The applicant stated that the design of non-Class 1 pressure vessels, heat exchangers, storage tanks, pumps, and turbine casings is typically in accordance with ASME Code Section VIII or

Section III, Subsection NC or ND (*i.e.*, Class 2 or 3). Some tank and pump designs meet other industry codes and standards (*e.g.*, American Water Works Association and Manufacturer's Standardization Society), reactor designer specifications, and architect engineer specifications. However, only ASME Code Section VIII, Division 2, and Section III, Subsection NC-3200 design codes include fatigue design requirements. The applicant stated that no detailed fatigue analyses are required due to the conservative requirements ASME Code Section VIII, Division 1, and Section III, Subsection NC-3100/ ND-3000. Cracking due to fatigue is not an aging effect requiring management for those components which do not require fatigue analysis. Fatigue analysis is not required for ASME Code Section VIII, Division I, Section III, Subsection NC-3100 or ND vessels nor for NC/ND pumps and storage tanks (less than 15 psig). The design specification indicates the applicable design code for each component. The applicant described fatigue TLAA dispositions in the following text only for the Unit 2 non-regenerative (letdown), regenerative, and RHR heat exchangers.

The applicant also stated that the Unit 2 non-regenerative (letdown) heat exchanger design complies with ASME Code Section III, Class C (tubes) and Section VIII, Division 1 (shell). Westinghouse Equipment Specification G-679150 defines the transients for the Unit 2 non-regenerative (letdown) heat exchanger. Its fatigue analysis is not bounding for 60 years of operation. The applicant will monitor the Unit 2 non-regenerative (letdown) heat exchanger transients with its Metal Fatigue of Reactor Coolant Pressure Boundary Program and will take corrective actions as appropriate (including reanalysis, repair, or replacement) to ensure that their design basis is not exceeded for the period of extended operation. Therefore, disposition of the Unit 2 non-regenerative (letdown) heat exchanger fatigue TLAA is in accordance with 10 CFR 54.21(c)(1)(iii).

The applicant further stated that the Unit 2 regenerative heat exchanger design complies with ASME Code Section III, Class 2. Westinghouse Equipment Specification G-679150 defines the transients for the Unit 2 regenerative heat exchanger. Its fatigue analysis is not bounding for 60 years of operation. The applicant will monitor the Unit 2 regenerative heat exchanger transients with its Metal Fatigue of Reactor Coolant Pressure Boundary Program and will take corrective actions as appropriate (including reanalysis, repair, or replacement) to ensure that the Unit 2 regenerative heat exchanger design basis is not exceeded for the period of extended operation. Therefore, disposition of the Unit 2 regenerative heat exchanger fatigue TLAA is in accordance with 10 CFR 54.21(c)(1)(iii).

The applicant also stated that the tube side design of the Unit 2 RHR heat exchangers complies with ASME Code Section III, Class 2, while the shell side design complies with ASME Code Section III, Class 3. Westinghouse Equipment Specification G-679150 defines transients applicable to these components. The fatigue analyses for the Unit 2 RHR heat exchangers are not bounding for 60 years of operation. The applicant will monitor the Unit 2 RHR heat exchanger transients with its Metal Fatigue of Reactor Coolant Pressure Boundary Program and will take corrective actions as appropriate (including reanalysis, repair, or replacement) to ensure that the Unit 2 RHR heat exchanger design basis is not exceeded for the period of extended operation. Therefore, disposition of the Unit 2 RHR heat exchanger fatigue TLAA is in accordance with 10 CFR 54.21(c)(1)(iii).

4.3.2.2.2 Staff Evaluation

The staff reviewed LRA Section 4.3.2.2 to verify, pursuant to 10 CFR 54.21(c)(1)(iii), that the effects of aging on the intended function(s) will be adequately managed for the period of extended operation.

In LRA Section 4.3.2.2, the applicant indicated that the Metal Fatigue of Reactor Coolant Pressure Boundary Program monitors the transients associated with non-regenerative (letdown) heat exchanger, regenerative heat exchanger, and RHR heat exchangers. However, LRA Section B.2.27 does not indicate that monitoring of the relevant transients will be provided by this AMP. In RAI B.2.27-10, dated May 28, 2008, that staff requested that the applicant (1) provide a list of the transients associated with the heat exchangers; (2) identify which of these transients are monitored by the program; and (3) explain its corrective actions when the current analyses are not bounding for 60 years of operation.

In its response to RAI B.2.27-10, dated July 11, 2008, the applicant clarified that all auxiliary system heat exchangers, which include letdown heat exchanger, regenerative heat exchanger and RHR heat exchangers, for both Unit 1 and Unit 2, are installed on the Class 2 part of the their respective systems and the primary side of these auxiliary heat exchangers are designed in accordance with ASME Section III, Class 2 requirements. The staff noted that since these heat exchangers were designed in accordance with ASME Section III, Class 2 rules, a fatigue analysis pursuant to ASME Section III Class 1 requirements is not applicable.

The staff noted that the expected total number of thermal cycles for the heat exchangers in question will be less than the 7000 thermal cycles required by ASME Class 2 thermal analysis; therefore, monitoring or a fatigue reanalysis is not required. By letter dated July 11, 2008, the applicant amended LRA Sections 4.3.2.2 and A.3.3.2.2 and the associated sub-sections and added LRA Section A.2.3.2.2 to reflect the discussion above. The staff noted that since these heat exchangers are bounded by 7000 equivalent full-temperature cycles for 60 years of operating, they will be no longer dispositioned under 10 CFR 54.21(c)(1)(iii), where the Metal Fatigue of Reactor Coolant Pressure Boundary Program is used for monitoring. The staff further noted that these heat exchangers will be dispositioned under 10 CFR 54.21(c)(1)(i), and that the TLAA remains valid for the period of extended operation.

Based on its review, the staff finds the applicant's response to RAI B.2.27-10 acceptable because the design of the heat exchangers in question is in compliance with ASME Code Section III, Class 2 rules, and the applicant has evaluated the heat exchangers to ensure that they will not exceed the 7000 equivalent full-temperature cycles. The staff concludes that the heat exchangers will not be monitored under the Metal Fatigue of Reactor Coolant Pressure Boundary Program and pursuant to 10 CFR 54.21(c)(1)(i), analyses will remain valid for the period of extended operation. Therefore, the staff's concern described in RAI B.2.27-10 is resolved.

4.3.2.2.3 UFSAR Supplement

The applicant provided a UFSAR supplement summary description of its TLAA evaluation of pressure vessels, heat exchangers, storage tanks, pumps, and turbine casings in LRA Section A.3.3.2.2. Based on its review of the UFSAR supplement, the staff concludes that the summary description of the applicant's actions to address pressure vessels, heat exchangers, storage tanks, pumps, and turbine casings is adequate.

4.3.2.2.4 Conclusion

Based on its review, as discussed above, the staff concludes that the applicant has demonstrated, pursuant to 10 CFR 54.21(c)(1)(i), that the analyses remain valid for the period of extended operation.

The staff also concludes that the UFSAR supplement contains an appropriate summary description of the TLAA evaluation, as required by 10 CFR 54.21(d).

4.3.3 Generic Industry Issues on Fatigue

This Section addresses both the applicant's fatigue TLAAs response to NRC Bulletins 88-08 and 88-11 and the effects of the primary coolant environment on fatigue life.

4.3.3.1 Thermal Stresses in Piping Connected to Reactor Coolant Systems (NRC Bulletin 88-08)

4.3.3.1.1 Summary of Technical Information in the Application

In LRA Section 4.3.3.1, the applicant summarized its evaluation of thermal stresses in RCS piping (NRC Bulletin 88-08, "Thermal Stresses in Piping Connected to Reactor Coolant Systems") for the period of extended operation. The applicant stated that NRC Bulletin 88-08 requested that licensees (1) review their RCS for any unisolable piping subject to temperature distributions which could result in unacceptable thermal stresses and any unisolable RCS piping sections that may have been subjected to excessive thermal stresses and (2) take action so such piping will not be subjected to unacceptable thermal stresses. There is no specific TLAA for Unit 1 and Unit 2 that responds to NRC Bulletin 88-08, except the Unit 2 RHR line analysis.

The applicant also stated that the Unit 2 RHR line stratification analysis required a detailed fatigue evaluation to demonstrate compliance with the design code of record (ASME Code Section III). Based on temperature data in response to NRC Bulletin 88-08, the applicant developed a conservative thermal stratification load case. Typical cycle periods for the thermal stratification events on the Unit 2 RHR lines were six to eight days, equating to approximately 2000 cycles for a 40-year plant life (assuming continuous stratification).
The fatigue analysis incorporated as an additional load, a bounding thermal stratification load which assumed 7000 cycles.

The applicant further stated that projection of the stratification cycles for a 60-year plant life results in 3000 cycles. The 7000 cycles in the fatigue analysis bounds the 60-year projected cycles; therefore, disposition of the Unit 2 RHR line fatigue TLAA is in accordance with 10 CFR 54.21(c)(1)(i).

4.3.3.1.2 Staff Evaluation

The staff reviewed LRA Section 4.3.3.1 to verify, pursuant to 10 CFR 54.21(c)(1)(i), that the analyses remain valid for the period of extended operation.

The staff reviewed the applicant's response to NRC Bulletin 88-08, (Letter to NRC, Beaver Valley Power Station, Unit No. 1, BV-1 Docket No. 50-334, License No. DPR-66, NRC Bulletin 88-08, February 7, 1990) and Letter to NRC, Beaver Valley Power Station, Unit No. 2, BV-2 Docket No. 50-412, License No. NPF-73, NRC Bulletin 88-08, July 14, 1989), in which the applicant stated that it will continue to monitor temperature until a long term solution is implemented to address the thermal stress in piping connected to RCS. In RAI 4.3-1, dated May 28, 2008, the staff requested that the applicant provide the follow-up actions taken in response to NRC Bulletin 88-08 and indicate whether temperature monitoring will be maintained to address thermal stratification during the period of extended operation.

In its response to RAI 4.3-1, dated July 11, 2008, the applicant provided details of its initial follow-up actions taken after NRC Bulletin 88-08 was issued, and further described the current and planned actions to address thermal stratification in piping connected to the RCS. The applicant stated that after NRC Bulletin 88-08 was issued, it began continuous temperature monitoring with thermocouples in February 1990 for Unit 1 and November 1989 for Unit 2. Monitoring was suspended in 2002 because the temporary instrumentation had become degraded and unreliable. Based on its review of the applicant's response letters to NRC Bulletin 88-08, the staff noted what appeared to be a discrepancy as to when the thermocouple monitoring began at Unit 1 and Unit 2. On August 28, 2008, the staff had a teleconference with the applicant to clarify the start date for thermocouple monitoring. The applicant, by letter dated October 2, 2008, clarified that the dates provided in its response to RAI 4.3-1 referred to the dates when data collection started, and not when data was obtained to establish a baseline. The staff noted that the applicant amended its original response to RAI 4.3-1, which stated that data collection to establish a baseline began in June 1989 for Unit 2 and December 1989 for Unit 1. The staff finds that the applicant has provided an appropriate clarification to the discrepancy in data collection start dates for Unit 1 and Unit 2.

The applicant collected initial data for both Unit 1 and Unit 2 in order to create a baseline temperature profile for each monitored line. Based on its review of the baseline data and subsequent data, the applicant showed that the temperature distribution did not fluctuate enough to create a large delta-temperature between the top and bottom of the pipe location. The applicant further described its actions, including weld inspections, that initially were performed for Unit 1 and Unit 2 during RFO 7 (September 1989) and RFO 1 (March 1989), respectively, and continued in subsequent RFOs. The applicant submitted a table indicating the date and the number of welds inspected during the RFO.

Based on the results of these weld inspections, the staff noted that no repairs were required. The staff further noted that these were the applicant's initial follow-up actions and that the applicant's current actions to address thermal stratification in piping connected to RCS must include its participation in Electric Power Research Institute initiatives that include the Thermal Stratification, Cycling and Striping project (Materials Reliability Program (MRP)-24) and MRP-146. The staff noted that all the piping lines within the scope of NRC Bulletin 88-08 and additional lines not originally within the scope of the NRC Bulletin 88-08 are in the scope of MRP-146 and have been screened. Those piping lines that have not been screened out, pursuant to the guidance in MRP-146, will include a detailed analysis.

The staff also noted that renewed thermocouple monitoring may be required for some of the piping lines, based on results from the detailed analysis. It was not clear to the staff whether the applicant had committed to thermocouple monitoring based on the detailed analysis; therefore, the staff held a teleconference with the applicant on August 28, 2008, in which the applicant stated that it will make a formal commitment to resume required thermocouple monitoring on those piping lines based on results from the detailed analysis. By letter dated October 2, 2008, the applicant committed (Commitments No. 31 and No. 32, for Unit 1 and Unit 2 respectively) to implement those actions that are needed, pursuant to the guidance found in MRP-146. The applicant stated that the "needed actions" for Unit 1 and Unit 2 include screening, detailed analysis, inspection, and temperature monitoring in accordance with the guidance provided in MRP-146. The applicant intends to perform augmented-nondestructive examination inspections during the next RFO at Unit 1 (Spring 2009) and Unit 2 (Fall 2009).

Based on its review of the applicant's response to RAI 4.3-1, the staff finds that the applicant has taken appropriate actions in response to NRC Bulletin 88-08 and will take appropriate actions to continue to address thermal stratification in piping lines connected to the RCS. The staff also finds that the applicant has committed (Commitments No. 31 and No. 32) to resume required thermocouple monitoring on those piping lines based on the detailed analysis performed pursuant to the guidance of MRP-146. Therefore, the staff's concern described in RAI 4.3-1 is resolved.

The staff noted that in LRA Section 4.3.3.1, the applicant stated, "Typical cycle periods for thermal stratification events on the Unit 2 RHR lines were six to eight days, which equated to approximately 2000 cycles for a 40-year plant life (assuming the stratification occurred continuously)." In RAI 4.3-7, dated May 28, 2008, the staff requested that the applicant provide the technical basis or its analyses which supports this statement.

In its response to RAI 4.3-7, dated July 11, 2008, the applicant stated that in its response to NRC Bulletin 88-08, Supplement 3, thermocouples were used to monitor the pipe temperatures for indication of thermal stratification for the Unit 2 RHR suction branch line. The staff confirmed in letter dated July 14, 1989, "Beaver Valley Power Station, Unit No. 2; Docket No. 50-412, License No. NPF-73 ; NRC Bulletin 88-08", and signed by J.D. Sieber, that the Unit 2 RHR suction branch line was instrumented with thermocouples to monitor for indication of thermal stratification and data has shown that stratification is occurring continuously. The applicant stated that based on the temperature data collected from the thermocouples an evaluation was performed which indicated that the typical cycle period for the thermal stratification was 6 to 8 days, assuming the stratification is occurring continuously.

The staff noted that based on the applicant's evaluation, this is equivalent to approximately 2000 cycles during the course of a 40-year plant life. The applicant stated that its fatigue analysis incorporated an additional load (thermal stratification load) that assumes 7000 cycles. Therefore, the applicant concluded that if the approximate 2000 cycles of stratification were projected to 60 years of operation, resulting in approximately 3000 cycles, the 7000 stratification cycles assumption for the Unit 2 RHR line fatigue analysis will remain valid for the period of extended operation.

Based on its review of the applicant's response to RAI 4.3-7 and the temperature data collected by the thermocouples, the staff finds that the applicant's analyses were based on a conservative approach. The staff further finds that the approximate 3000 stratification cycles projected to 60 years of operation is bounded by the 7000 cycles assumed in the fatigue analyses and, remains valid for the period of extended operation in accordance with 10 CFR 54.21(c)(1)(i). Therefore, the staff's concern described in RAI 4.3-7 is resolved.

4.3.3.1.3 UFSAR Supplement

The applicant provided a UFSAR supplement summary description of its TLAA evaluation of thermal stresses in piping connected to RCSs (NRC Bulletin 88-08) in LRA Section A.3.3.3.1. Based on its review of the UFSAR supplement, the staff concludes that the summary description of the applicant's actions to address thermal stresses in piping connected to RCSs (NRC Bulletin 88-08) is adequate.

4.3.3.1.4 Conclusion

Based on its review, as discussed above, the staff concludes that the applicant has demonstrated, pursuant to 10 CFR 54.21(c)(1)(i), that, for thermal stresses in piping connected to RCSs (NRC Bulletin 88-08), the analyses remain valid for the period of extended operation. The staff also concludes that the UFSAR supplement contains an appropriate summary description of the TLAA evaluation, as required by 10 CFR 54.21(d).

4.3.3.2 *Pressurizer Surge Line Thermal Stratification (NRC Bulletin 88-11)*

4.3.3.2.1 Summary of Technical Information in the Application

In LRA Section 4.3.3.2, the applicant summarized its evaluation of pressurizer surge line thermal stratification for the period of extended operation. NRC Bulletin 88-11, "Pressurizer Surge Line Thermal Stratification," requires a plant-specific or generic demonstration that the pressurizer surge line meets design code requirements for the effects of thermal stratification.

BVPS Unit 1 Evaluation. The applicant stated that it had participated in a WOG program for partial resolution of this issue. The program collected, summarized, and evaluated pressurizer surge line physical and operating data relating to piping layout, supports and restraints, components, size, material, operating history, etc., for all domestic Westinghouse PWRs in conjunction with available monitoring data and plant-specific analyses by Westinghouse.

The applicant also stated that in January 1991, it had submitted and the staff approved WCAP-12727, "Evaluation of Thermal Stratification for the Beaver Valley Unit 1 Pressurizer Surge Line." The applicant reviewed WCAP-12727 for impact resulting from the extended power uprate and performed a detailed analysis at the controlling location (reactor coolant loop (RCL) nozzle) to account for temperature effects of the power uprate. The applicant then calculated a new CUF demonstrated to remain less than the code-allowable limit of 1.0.

BVPS Unit 2 Evaluation. The applicant stated that it had first observed apparent surge line stratification during the Unit 2 hot functional testing which preceded NRC Bulletin 88-11. Based on this observation, the applicant revised its surge line ASME Code Section III analysis of record to evaluate stress and fatigue effects with data from additional instrumentation temporarily installed to monitor pipe and fluid conditions. Subsequently, the applicant contracted with Westinghouse for a complete reanalysis of surge line thermal stratification and striping.

The staff accepted WCAP-12093, "Evaluation of Thermal Stratification for the Beaver Valley Unit 2 Pressurizer Surge Line," as meeting leak-before-break (LBB) requirements and other NRC Bulletin 88-11 concerns for the Unit 2 surge line, and demonstrating that thermal stratification effects do not cause the pressurizer surge line to exceed code-allowable design limits.

The applicant reviewed WCAP-12093 for impact resulting from the extended power uprate and performed a detailed analysis at the controlling location (RCL nozzle) to account for temperature effects of the power uprate. The applicant then calculated a new CUF demonstrated to remain less than the code-allowable limit of 1.0.

BVPS Units 1 and 2 Disposition for License Renewal. The applicant stated that both WCAP-12727 and WCAP-12093 determine the effect of thermal stratification by imposing defined thermal stratification cycles upon the stress and fatigue evaluations. The stratification cycles incorporated into the CUF determination are defined by the 200 heatup and cooldown design transients; therefore, these NRC Bulletin 88-11 analyses are TLAAs in accordance with 10 CFR 54.3. LRA Section 4.3.4 demonstrates that the 200 heatup and cooldown cycles are bounding for 60 years of operation; therefore, disposition of the Unit 1 and Unit 2 pressurizer surge line fatigue TLAAs are in accordance with 10 CFR 54.21(c)(1)(i).

4.3.3.2.2 Staff Evaluation

The staff reviewed LRA Section 4.3.3.2 to verify, pursuant to 10 CFR 54.21(c)(1)(i), that the analyses remain valid for the period of extended operation.

In LRA Section 4.3.3.2.2, the applicant stated that it had performed a detailed analysis at the controlling location (reactor coolant loop (RCL) nozzle), to account for defined thermal stratification and temperature effects due to the thermal power uprate. Those analyses, which supplemented the applicant's original analyses (WCAP-12727 for Unit 1 and WCAP-12093 for Unit 2), demonstrated the new CUF remain less than the code-allowable limit of 1.0. The staff noted that in LRA Section 4.3.3.2, the applicant stated that those analyses remain valid for the license renewal period and were dispositioned it in accordance with 10 CFR 54.21(c)(1)(i). In RAI 4.3-6, dated May 28, 2008, the staff requested that the applicant provide the basis for this statement.

In its response to RAI 4.3-6, dated July 11, 2008, the applicant amended LRA Sections 4.3.3.2.3, A.2.3.3.1 and A.3.3.3.2 to reflect the pressurizer surge line as dispositioned in accordance with 10 CFR 54.21(c)(1)(iii) rather than 10 CFR 54.21(c)(1)(i). The applicant explained that because it used the 60-year projected operational cycles to determine that the 200 heatup and cooldown transients remain valid for the period of extended operation, the Metal Fatigue of Reactor Coolant Pressure Boundary Program for Unit 1 and Unit 2 must be used to validate this same assumption.

Based on its review, the staff finds the applicant's response to RAI 4.3-6 acceptable because the applicant (1) has amended the LRA to disposition the pressurizer surge line in accordance with 10 CFR 54.21(c)(1)(iii); (2) will monitor this location with the Metal Fatigue of Reactor Coolant Pressure Boundary Program and; (3) will initiate appropriate corrective actions so that the effects of aging on the intended functions of these components are adequately managed for the period of extended operation. Therefore, the staff's concern described in RAI 4.3-6 is resolved.

4.3.3.2.3 UFSAR Supplement

The applicant provided a UFSAR supplement summary description of its TLAA evaluation of pressurizer surge line thermal stratification (NRC Bulletin 88-11) in LRA Sections A.2.3.3.1 and A.3.3.3.2. Based on its review of the UFSAR supplement, the staff concludes that the summary description of the applicant's actions to address pressurizer surge line thermal stratification (NRC Bulletin 88-11) is adequate.

4.3.3.2.4 Conclusion

Based on its review, as discussed above, the staff concludes that the applicant has demonstrated, pursuant to 10 CFR 54.21(c)(1)(iii), that, for pressurizer surge line thermal stratification (NRC Bulletin 88-11), the effects of aging on the intended function(s) will be adequately managed for the period of extended operation. The staff also concludes that the UFSAR supplement contains an appropriate summary description of the TLAA evaluation, as required by 10 CFR 54.21(d).

4.3.3.3 Effects of Primary Coolant Environment on Fatigue Life

4.3.3.3.1 Summary of Technical Information in the Application

In LRA Section 4.3.3.3, the applicant summarized its evaluation of the effects of primary coolant environment on fatigue life for the period of extended operation. Test data indicate that certain environmental conditions (*e.g.*, temperature, oxygen content, strain rate) in the primary systems of light water reactors could cause greater susceptibility to fatigue than would be predicted by fatigue analyses based on the ASME Code Section III design fatigue curves from laboratory tests in air and at low temperatures. Adjustments to failure curves derived from laboratory tests to account for data scatter, size effect, and surface finish may not be sufficient to account for actual plant operating environments.

Study of environmental effects on the fatigue life of selected components was under two generic issues, Generic Safety Issue (GSI)-78, "Monitoring of Fatigue Transient Limits for Reactor Coolant System," and GSI-166, "Adequacy of Fatigue Life of Metal Components." GSI-78, determined whether fatigue monitoring was necessary at operating plants and calculated risk from through-wall cracking of metal components due to fatigue. GSI-166 assessed the significance of more recent fatigue test data on the fatigue life of a sample of components in plants that had analyzed code fatigue design. A fatigue action plan coordinated efforts on fatigue life estimation and addressed ongoing GSI-78 and GSI-166 issues for 40-year plant life.

In closing GSI-166, the staff concluded that the environmental effects on fatigue life are not safety-related through the end of the initial license term. This conclusion was based on two studies. The first, published as NUREG/CR-6260, "Application of NUREG/CR-5999 Interim Fatigue Curves to Selected Nuclear Power Plant Components," applied the fatigue design curves for environmental effects to several plants and specified locations of interest for consideration. The second study, based on a risk analysis on fatigue failures, concluded that environmental effects on core damage frequency are insignificant. These two studies concluded that environmental effects are not a concern for the current license term. Closure of GSI-166 led to GSI-190, "Fatigue Evaluation of Metal Components for 60-Year Plant Life." In closing GSI-190 on the effects of a reactor water environment on fatigue life, the staff concluded licensees should address the effects of the coolant environment on component fatigue life as they formulate AMPs for license renewal.

In summary, the staff concluded that environmental effects have a negligible impact on core damage frequency and therefore require no generic regulatory action but that environmental effects can increase the frequency of pipe leaks and that applicants for license renewal should address the effects of reactor coolant environment on component fatigue life in their aging management reviews.

The applicant's management of the environmental effects upon component fatigue life determines limiting locations based on the NRC-sponsored studies reported in NUREG/CR-6260 for reevaluation guided by NUREG-1801, "Generic Aging Lessons Learned (GALL) Report," Section X.M1, to demonstrate maintenance of CUFs at such locations below the code-allowable limit of 1.0.

BVPS Units 1 and 2 NUREG/CR-6260 Location Determination. NUREG/CR-6260 applies the fatigue design curves for environmental effects to several plant designs. As Unit 1 and Unit 2 were designed at different times, the plants are different vintages of Westinghouse-designed plants based on the RCS design code. The Unit 1 RCPB piping design is to ANSI B31.1, and Unit 1 is therefore an older-vintage Westinghouse plant. The RCPB piping for Unit 2 is designed to ASME Code Section III, and Unit 2 is therefore a newer-vintage Westinghouse plant.

Section 5.5 of NUREG/CR-6260 specifies the following component locations as representative for environmental effects for older-vintage Westinghouse plants. These locations and the subsequent calculations directly relevant to Unit 1 are as follows:

- Reactor vessel shell and lower head (shell-to-head transition)
- Reactor vessel inlet and outlet nozzles
- Pressurizer surge line (hot leg nozzle safe end)
- RCS piping charging system nozzle
- RCS piping safety injection nozzle
- RHR system tee

Section 5.4 of NUREG/CR-6260 specifies the following component locations as representative for environmental effects for newer-vintage Westinghouse plants. These locations and the subsequent calculations directly relevant to Unit 2 are as follows:

- Reactor vessel shell and lower head (shell-to-head transition)
- Reactor vessel inlet and outlet nozzles
- Pressurizer surge line (hot leg nozzle safe end)
- RCS piping charging system nozzle (knuckle region)
- RCS piping safety injection nozzle (knuckle region)
- RHR system piping (inlet piping transition)

BVPS Units 1 and 2 NUREG/CR-6260 Location Environmental Fatigue Evaluation. The applicant's evaluation of Unit 1 and Unit 2 NUREG/CR-6260 locations used the guidance of NUREG/CR-6583, "Effects of LWR Coolant Environments on Fatigue Design Curves of Carbon and Low Alloy Steels," and NUREG/CR-5704, "Effects of LWR Coolant Environments on Fatigue Design Curves of Austenitic Stainless Steels." These reports describe the use of a fatigue life correction factor (F_{en}) to express the effects of the reactor coolant environment upon the material fatigue life. Determination of the expression for F_{en} was through experimental and statistical data. F_{en} for carbon and low alloy steel is a function of fluid service temperature, material sulfur content, fluid-dissolved oxygen, and strain rate. For austenitic stainless steel, F_{en} is a function of fluid service temperature, fluid-dissolved oxygen, and strain rate. Determination of the CUF, including environmental effects (U_{env}), is from the existing 60-year CUF (U_{60}) through the use of the F_{en}:

$$U_{env} = U_{60} * F_{en}$$

In order for the applicant to demonstrate acceptable fatigue life including environmental effects, the CUF, including environmental effects, should remain less than design code-allowable (*i.e.*, U_{env} 1.0). Therefore, the applicant applied F_{en} to the CUFs at the Unit 1 and Unit 2 NUREG/CR-6260 locations and compared the results to the design code-allowable limit. It should be noted that three of the Unit 1 NUREG/CR-6260 locations (charging system nozzle, safety injection nozzle, and the RHR system tee) are designed to the ANSI B31.1 standard, which does not require determination of usage factors for fatigue evaluations. Therefore, re-evaluation of these locations in accordance with ASME Code Section III, 1989 Edition with 1989 Addenda, determined 60-year CUFs, applied the appropriate F_{en} to these CUFs, and compared the results against the ASME Code Section III allowable limit. In LRA Table 4.3-1, the applicant provided detailed results of its evaluations of environmental fatigue.

BVPS Units 1 and 2 Disposition for License Renewal. At several locations (Unit 1 pressurizer surge line and charging system nozzle, Unit 2 pressurizer surge line, charging system nozzle, and RHR system piping), U_{env} exceeded the 1.0 design code-allowable limit. For these locations, the applicant will implement one or more of the following as required by the Metal Fatigue of Reactor Coolant Pressure Boundary Program:

(1) Further refinement of the fatigue analyses to lower the predicted CUFs to less than 1.0

(2) Management of fatigue at the affected locations by an inspection program reviewed and approved by the staff (*e.g.*, inspection intervals to be determined by an acceptable method)

(3) Repair or replacement of the affected locations

If the applicant opts to manage environmental-assisted fatigue during the period of extended operation, it will submit AMP details (scope, qualification, method, and frequency) to the staff prior to the period of extended operation; therefore, the applicant dispositioned the TLAAs for the Unit 1 pressurizer surge line and charging system nozzle and the Unit 2 pressurizer surge line, charging system nozzle, and RHR system piping in accordance with 10 CFR 54.21(c)(1)(iii).

The CUFs, including environmental fatigue at the other limiting locations (Unit 1 RV shell and lower head, RV inlet and outlet nozzles, safety injection nozzle and RHR system tee; Unit 2 RV shell and lower head, RV inlet and outlet nozzles, and safety injection nozzle) remain demonstrably less than the 1.0 design code-allowable limit for the period of extended operation; therefore, the applicant dispositioned the TLAAs for these other locations in accordance with 10 CFR 54.21(c)(1)(ii).

4.3.3.3.2 Staff Evaluation

The staff reviewed LRA Section 4.3.3.3 to verify, pursuant to 10 CFR 54.21(c)(1)(ii), that the analyses have been projected to the end of the period of extended operation and, pursuant to 10 CFR 54.21(c)(1)(iii), that the effects of aging on the intended function(s) will be adequately managed for the period of extended operation.

In LRA Section 4.3.3.3, the applicant discussed the effects of primary coolant environment on fatigue life. During the audit, the applicant indicated that it will refine its analysis for NUREG/CR-6260 components in the near future. To assist its review, the staff issued RAI 4.3-3, dated May 28, 2008, requesting that the applicant (1) provide the schedule for refining its analysis for the environmental-assisted fatigue of the NUREG/CR-6260 locations in which the CUF, including environmental effects (U_{env}), exceeded the design code allowable value; (2) explain how the calculations for the F_{en}, used to express the effects of the reactor coolant environment, will be performed and specifically, how the transient pairs will be considered in the calculations; and (3) describe the criteria and methodology that will be used to perform the additional analyses in calculating the CUF, including U_{env}, for components that exceed the design code-allowable value of 1.0.

In its response to RAI 4.3-3, dated July 11, 2008, the applicant provided a schedule for the reanalysis of those components where U_{env} exceeded the design code-allowable limit of 1.0. Furthermore, the applicant committed (Regulatory Commitment No. 1) to perform this reanalysis and to submit its results to the staff, along with a summary of how the analysis was performed, no later than October 15, 2008.

The staff noted in the applicant's response to RAI 4.3-3 that the Unit 1 and Unit 2 surge line to hot leg nozzle and charging system nozzle, and the Unit 2 safety injection system nozzle and RHR system piping are all fabricated of stainless steel. The applicant stated that the general methodology it used to calculate the F_{en} was the guidance found in NUREG/CR-5704. The applicant further explained that the fatigue usage is calculated with F_{en} factors applied on each load pair incremental usage for the Unit 1 and Unit 2 surge line to hot leg nozzle only, whereas the bounding F_{en} factor is applied to the design CUF for the remaining locations. The applicant expects that results from the refined reanalysis will be successfully based on the methodology provided in the response for all the locations mentioned above; however, as an alternative analysis for the Unit 1 surge line to hot leg nozzle, the applicant may perform a fracture mechanics analysis in accordance with the general methodology described in NUREG/CR-6934.

NUREG/CR-6934, as noted by the staff, is not endorsed by the NRC. Therefore, the staff held a teleconference with the applicant on September 4, 2008, during which time the staff explained that NUREG/CR-6934 is not endorsed by the NRC and thus the results of the applicant's reanalysis are subject to staff review and approval. By letter dated October 2, 2008, the applicant acknowledged the staff's concern and stated that it had completed the reanalysis of those locations listed in Regulatory Commitment No. 1. The applicant further stated that the CUF included environmental factors from the reanalysis of the Unit 1 and Unit 2 charging system nozzle and determined that the Unit 2 safety injection nozzle and RHR system piping will remain below the code-allowable limit of 1.0 during the period of extended operation. However, for the Unit 1 and Unit 2 pressurizer surge line to hot leg nozzle, the applicant stated that the CUF, including environmental factors, exceeded the code-allowable limit of 1.0. The applicant further stated that the Unit 1 and Unit 2 pressurizer surge line to hot leg nozzle will be managed by the Metal Fatigue of Reactor Coolant Pressure Boundary Program and is within the scope of Commitments No. 25 and No. 26, for Unit 1 and Unit 2 respectively.

The applicant has since withdrawn Regulatory Commitment No. 1 and the proposed use of NUREG/CR-6934 because the applicant has completed its analysis and has placed the Unit 1 and Unit 2 pressurizer surge line to hot leg nozzle within the scope of the Metal Fatigue of Reactor Coolant Pressure Boundary Program. The staff confirmed that the applicant has amended LRA Sections 4.3.3.3.3, Section A.2.3.3.2 and Section A.3.3.3.3 to state that the Metal Fatigue of Reactor Coolant Pressure Boundary Program will manage all NUREG/CR-6260 locations because the 60-year projected operational cycles were used in the design fatigue analysis and will require validation of the assumptions used in the analysis.

On the basis of its review, the staff finds the applicant's response to RAI 4.3-3 acceptable because (1) the applicant has completed the reanalysis and provided the results and methodology in letter dated October 2, 2008, which demonstrated that the CUF, including environmental factors for the NUREG/CR-6260 locations, will remain below the code-allowable limit of 1.0, except for the Unit 1 and Unit 2 pressurizer surge line to hot leg nozzle; (2) the applicant will manage the all NUREG/CR-6260 locations, including the Unit 1 and Unit 2 pressurizer surge line to hot leg nozzle, with the Metal Fatigue of Reactor Coolant Pressure Boundary Program; and (3) the applicant had calculated the F_{en} factor for those locations requiring reanalysis for stainless steels in accordance with NUREG/CR-5704. Therefore, the staff's concern described in RAI 4.3-3 is resolved.

During the audit, the staff noted in LRA Table 4.3-1 that the 60-year CUF (U_{60}) value as well as the environmental CUF value for the Unit 2 safety injection system is not correct. The staff considered the deletion of the boron injection tank line for Unit 2 and confirmed that these values in LRA Table 4.3-1 do not represent the results for safety injection nozzle to the cold leg. In RAI 4.3-13, dated May 28, 2008, the staff requested that the applicant provide the 60-year CUFs (U_{60}) and environmental-assisted fatigue results for this location as recommended by NUREG/CR-6260.

In its response to RAI 4.3-13, dated July 11, 2008, the applicant stated that this discovery made during the audit is currently being addressed with the FENOC corrective actions program. The staff noted that if the applicant used the design CUF from the correct location, taking into account environmental-assisted fatigue, the usage factor would exceed the code-allowable limit of 1.0. The applicant stated that a reanalysis is required for this NUREG/CR-6260 location and committed (Regulatory Commitment No. 1) to perform the reanalysis for the applicable NUREG/CR-6260 locations, including the safety injection nozzle; and, to submit its results to the staff, along with a summary of how the analysis was performed, no later than October 15, 2008.

On the basis of its review, the staff finds the applicant's response to RAI 4.3-13 acceptable because the applicant has completed its reanalysis and provided the results and methodology to the staff by letter dated October 2, 2008. The staff also finds that the applicant has demonstrated that the CUF, including environmental factors for the NUREG/CR-6260 locations, will remain below the code-allowable limit of 1.0, except for the Unit 1 and Unit 2 pressurizer surge line to hot leg nozzle; and that the applicant will manage all NUREG/CR-6260 locations with the Metal Fatigue of Reactor Coolant Pressure Boundary Program. Therefore, the staff's concern described in RAI 4.3-13 is resolved.

In LRA Section 4.3.3.3.2, the applicant stated that three of the NUREG/CR-6260 locations for Unit 1 were re-evaluated in accordance with ASME Section III, 1989 Edition with 1989 addenda to determine the 60-year CUFs. The staff noted that the applicant performed its analysis on the Unit 1 charging nozzle, safety injection nozzle and the RHR system tee in accordance with ASME Section III, 1989 Edition with 1989 addenda because these locations were originally designed pursuant to ANSI B31.1 standards. As a result, usage factors were not determined as part of the fatigue evaluation. In RAI 4.3-15, dated May 28, 2008, the staff requested that the applicant provide the design basis transients and the associated cycles used to calculate the 60-year CUFs (U_{60}) in LRA Table 4.3-1.

In its response to RAI 4.3-15, dated July 11, 2008, the applicant provided the information requested by the staff for the Unit 1 charging nozzle, safety injection nozzle and the RHR system tee. The staff noted that the U_{env} for the Unit 1 charging nozzle exceeded the code-allowable limit of 1.0 when considering environmentally assisted fatigue. The applicant committed (Regulatory Commitment No. 1) to perform a reanalysis for this location. As part of this commitment, the applicant will complete the reanalysis and submit the results to the staff, along with a summary of how the analysis was performed, no later than October 15, 2008. The staff finds this portion of the applicant's response acceptable.

By letter dated October 2, 2008, the applicant submitted the results of its reanalysis of the Unit 1 charging nozzle to the staff. The response detailed the applicant's methodology which demonstrated that the CUF, including environmental factors for the NUREG/CR-6260 locations, will remain below the code-allowable limit of 1.0, except for the Unit 1 and Unit 2 pressurizer surge line to hot leg nozzle. The applicant stated that it will manage all NUREG/CR-6260 locations with the Metal Fatigue of Reactor Coolant Pressure Boundary Program and that it had calculated the F_{en} factor for those locations requiring reanalysis for stainless steels in accordance with NUREG/CR-5704.

The applicant further stated in its response to RAI 4.3-15 that it used the applicable design transients from the general piping analysis for Unit 2 for the re-evaluation of the Unit 1 safety injection nozzle and the RHR system tee, in accordance with ASME Section III. The staff noted that Unit 1 was designed pursuant to ANSI B31.1 standards and Unit 2 was designed in accordance with ASME Code Section III. The staff also noted that the applicant utilized the applicable design transients from the general piping analysis from Unit 2 and the design cycles of these transients because of the similarity of design and operation between both units, thus making Unit 2 representative of Unit 1. The staff further noted that there was one exception, the design transient "RHR operation", in which the applicant had increased the design cycles in the fatigue analysis to account for the projected cycles. The staff finds the applicant's response acceptable because the applicant has provided adequate information to the staff explaining which design transients and the number of design cycles were used in its fatigue analysis for the Unit 1 safety injection nozzle and RHR system tee.

Based on its review, the staff finds that the applicant's response to RAI 4.3-15 acceptable because the applicant has committed (Regulatory Commitment No. 1) to reanalyze the Unit 1 charging nozzle and to submit the results and methodology used for the analysis to the staff, no later than October 15, 2008. The staff further finds, by letter dated October 2, 2008, that the applicant has completed its reanalysis of the Unit 1 charging nozzle and has provided the results and methodology which demonstrate that the CUF, including environmental factors for the NUREG/CR-6260 locations, will remain below the code-allowable limit of 1.0, except for the Unit 1 and Unit 2 pressurizer surge line to hot leg nozzle. The staff also finds that the applicant will manage all NUREG/CR-6260 locations with the Metal Fatigue of Reactor Coolant Pressure Boundary Program and has calculated the F_{en} factor for those locations that required reanalysis for stainless steels in accordance with NUREG/CR-5704. Therefore, the staff's concern described in RAI 4.3-15 is resolved.

The staff noted that the 60-year CUF (U_{60}) for the Unit 2 RHR system piping in LRA Table 4.3-1 is higher than for Unit 1 and that the Unit 2 RHR system piping is dispositioned pursuant to 10 CFR54.21(c)(1)(iii). During the audit, the applicant indicated that the analysis for the 60-year CUF (U_{60}) will be refined. In RAI 4.3-8, dated May 28, 2008, the staff requested that the applicant explain in detail how the RHR system piping will be managed for aging effects.

In its response to RAI 4.3-8, dated July 11, 2008, the applicant committed (Regulatory Commitment No. 1) to perform a reanalysis for the applicable NUREG/CR-6260 locations, including the Unit 2 RHR system piping, and to submit the results of the reanalysis to the staff, along with a summary of how the analysis was performed, no later than October 15, 2008. The applicant further explained that since Unit 1 and Unit 2 are not the same vintage Westinghouse design, the results of the CUF are not directly comparable.

Based on its review, the staff finds the applicant's response to RAIs 4.3-8 acceptable because the applicant has committed (Regulatory Commitment No. 1) to reanalyze the Unit 2 RHR system piping and to submit the results and methodology used for the analysis to the staff, no later than October 15, 2008. The staff further finds, by letter dated October 2, 2008, that the applicant has completed its reanalysis of the Unit 2 RHR system piping and has provided the results and methodology which demonstrate that the CUF, including environmental factors for the NUREG/CR-6260 locations, will remain below the code-allowable limit of 1.0, except for the Unit 1 and Unit 2 pressurizer surge line to hot leg nozzle. The staff also finds that the applicant will manage all NUREG/CR-6260 locations with the Metal Fatigue of Reactor Coolant Pressure Boundary Program and has calculated the F_{en} factor for those locations that required reanalysis for stainless steels in accordance with NUREG/CR-5704. Therefore, the staff's concern described in RAI 4.3-8 is resolved.

In LRA Table 4.3-1 and Section 4.3.3.3, the applicant provided the TLAA disposition for Unit 1 and Unit 2 to address environmental assisted fatigue. The staff noted the TLAAs for some of the locations appeared to be dispositioned pursuant to 10 CFR 54.21(c)(1)(i), but in LRA Section 4.3.3.3, the applicant indicated that these components were dispositioned in accordance with 10 CFR 54.21(c)(1)(ii). In RAI 4.3-9, dated May 28, 2008, the staff requested that the applicant clarify the TLAA dispositions for the each of the NUREG/CR-6260 locations.

In its response to RAI 4.3-9, dated July 11, 2008, the applicant explained that it used the 60-year projected operational cycles in the fatigue evaluations for those locations, where U_{env} has been demonstrated to remain below the code-allowable limit of 1.0 (Unit 1 RV shell and lower head, RV inlet and outlet nozzles, safety injection nozzle and RHR system tee; Unit 2 RV shell and lower head, RV inlet and outlet nozzles) for the period of extended operation. The applicant stated that the Metal Fatigue of Reactor Coolant Pressure Boundary Program must be used to validate the assumptions used in these evaluations and amended LRA Section 4.3.3.3.3 to read that for Unit 1 and Unit 2, all locations recommended by NUREG/CR-6260 will be dispositioned in accordance with 10 CFR 54.21(c)(1)(iii). The applicant further stated that it will reanalyze NUREG/CR-6260 locations in which U_{env} exceeded the code-allowable limit of 1.0; and, committed (Regulatory Commitment No. 1) to perform this reanalysis and submit the results to the staff, along with a summary of how the analysis was performed, no later than October 15, 2008.

By letter dated October 2, 2008 the applicant has (1) completed the reanalysis and provided the results and methodology which demonstrated that the CUF including environmental factors for the NUREG/CR-6260 locations will remain below the code allowable limit of 1.0 except for the Unit 1 and Unit 2 pressurizer surge line to hot leg nozzle (2) the applicant will manage all NUREG/CR-6260 locations with the Metal Fatigue of Reactor Coolant Pressure Boundary Program and (3) the applicant calculated the F_{en} factor for those locations that required reanalysis for stainless steels in accordance with NUREG/CR-5704.

Based on its review, the staff finds the applicant's response to RAIs 4.3-9 acceptable because the applicant has dispositioned all of the NUREG/CR-6260 locations in accordance with 10 CFR 54.21(c)(1)(iii) and will use the Metal Fatigue of Reactor Coolant Pressure Boundary Program to adequately manage the aging of these components for the period of extended operation. The staff further finds that the applicant has committed (Regulatory Commitment No. 1) to reanalyze the NUREG/CR-6260 locations in which U_{env} exceeded the code-allowable limit of 1.0. Unit 2 RHR system piping and to submit the results and methodology used for the analysis to the staff, no later than October 15, 2008.

The staff also finds, by letter dated October 2, 2008, that the applicant has completed its reanalysis of the NUREG/CR-6260 locations and has provided the results and methodology which demonstrate that the CUF, including environmental factors for the NUREG/CR-6260 locations, will remain below the code-allowable limit of 1.0, except for the Unit 1 and Unit 2 pressurizer surge line to hot leg nozzle.

In addition, the staff finds that the applicant will manage all NUREG/CR-6260 locations with the Metal Fatigue of Reactor Coolant Pressure Boundary Program and has calculated the F_{en} factor for those locations that required reanalysis for stainless steels in accordance with NUREG/CR-5704. Therefore, the staff's concern described in RAI 4.3-9 is resolved.

4.3.3.3.3 UFSAR Supplement

The applicant provided a UFSAR supplement summary description of its TLAA evaluation of effects of the primary coolant environment on fatigue life in LRA Sections A.2.3.3.2 and A.3.3.3.3. Based on its review of the UFSAR supplement, the staff concludes that the summary description of the applicant's actions to address the effects of primary coolant environment on fatigue life is adequate.

4.3.3.3.4 Conclusion

Based on its review, as discussed above, the staff concludes that the applicant has demonstrated, pursuant to 10 CFR 54.21(c)(1)(iii), that for effects of the primary coolant environment on fatigue life, the effects of aging on the intended function(s) will be adequately managed for the period of extended operation. The applicant also has demonstrated, pursuant to 10 CFR 54.21(c)(1)(iii), that the effects of aging on the intended function(s) will be adequately managed for the period of extended operation.

The staff also concludes that the UFSAR supplement contains an appropriate summary description of the TLAA evaluation, as required by 10 CFR 54.21(d).

4.3.4 Nuclear Steam Supply System Transient Cycle Projection For 60-Year Operation

4.3.4.1 Summary of Technical Information in the Application

In LRA Section 4.3.4, the applicant summarized its evaluation of NSSS transient cycle projection for the 60-year period of extended operation. The applicant indicated the transients it used for calculating fatigue usage factors for the NSSS. For this set of cyclic design transients, the applicant compiled the number of operational cycles accrued to October 2003 and projected the number at the end of 60 years of operation to determine whether the results remain below the number of design-allowable cycles.

The applicant also stated that it had extrapolated the number of transients to be accumulated by 60 years of operation. The two options for extrapolating the number of transient cycles are:

(1) Develop histograms of each transient and, based on recent operating history (*i.e.*, the last ten years), project the cumulative number of operational cycles at 60 years

(2) Linearly extrapolate the cumulative number of operational cycles at 60 years

Because plant performance has improved with time, the first option typically results in a more accurate projection, the second, in a more conservative number of thermal cycles. Except for the plant heatup and cooldown, pressurizer cooldown, and reactor trip transients, the extrapolation for all transients used the second option. For the plant heatup and cooldown and for pressurizer cooldown, the projection of cycles used the first option. The applicant also chose the first option for the reactor trip transients but biased the extrapolation with additional reactor trips as the unit approaches end of life (EOL). Accrued operational cycles based on initial operations for Unit 1 of 1975 and Unit 2 of 1986 use a current plant life as of October 2003; therefore, the operating lifetimes for the evaluations were 28 and 17 years for Unit 1 and Unit 2, respectively. LRA Table 4.3-2 presents the results of the transient cycle extrapolation.

4.3.4.2 Staff Evaluation

The staff reviewed LRA Section 4.3.4, pursuant to 10 CFR 54.21(c)(1).

In LRA Section 4.3.4, the applicant states that histograms were developed based on the last ten years in order to perform an extrapolation for the number of accumulated transients in 60 years of operation for plant heatup and cooldown, and pressurizer cooldown. In RAI 4.3-16, dated

4-50

May 28, 2008, the staff requested that the applicant provide the histograms that were developed and the method used by the applicant to extrapolate these cycles to 60 years of operation.

In its response to RAI 4.3-16, dated July 11, 2008, the applicant provided the staff with the histograms for the Unit 1 heatup, cooldown and reactor trip projection. The applicant also provided the methodology that it used to extrapolate the projection to 60 years for these transients. The applicant explained that the histograms incorporated the Unit 1 heatup, cooldown and reactor trip transient cycles accrued through May 1, 2007 and are all subject to bias with additional cycles as Unit 1 approaches the EOL (60 years of operation). The staff noted that since these transients are expected to approach or exceed the number of design cycles during the period of extended operation, the applicant's Metal Fatigue of Reactor Coolant Pressure Boundary Program is required to monitor these transients and to initiate corrective actions if the triggering point is reached. The staff further noted that in the response, the applicant did not provide the histogram for the pressurizer cooldown. Therefore, on August 28, 2008, the staff held a teleconference with the applicant, during which time the applicant explained that the transient "pressurizer cooldown" is not independently tracked; and, therefore a histogram does not exist. By letter dated October 2, 2008, the applicant amended the LRA to remove the reference to the pressurizer cooldown transient since it is not applicable to Unit 1 and Unit 2.

Based on its review, the staff finds the applicant's response to RAI 4.3-16 acceptable because (1) the applicant provided the applicable histograms that were requested, which conservatively biased additional cycles of each transient as Unit 1 approaches the EOL (60 years of operation); (2) the applicant will monitor these transients as part of its Metal Fatigue of Reactor Coolant Pressure Boundary Program in accordance with 10 CFR 54.21(c)(1)(iii); and (3) that the effects of aging on the intended function(s) will be adequately managed for the period of extended operation. Therefore, the staff's concern described in RAI 4.3-16 is resolved.

In LRA Table 4.3-2, the staff noted that the 60-year projected operational cycle for operating-basis earthquakes (OBEs) is not provided. In RAI 4.3-11, dated May 28, 2008, the staff requested that the applicant confirm the number of OBE occurrences or stress cycles it will consider in the 60-year EAF evaluation.

In its response to RAI 4.3-11, dated July 11, 2008, the applicant stated that based on its response to RAI 4.3-3, the analyses for environmentally assisted fatigue is still on going. The applicant further stated that for those analyses that have already been completed, a minimum of 50 OBE cycles have been incorporated. However, for those analyses that have not yet been completed, the applicant also intends to use a minimum of 50 cycles of OBE. If fewer cycles are used, the applicant will report this change to the staff, along with the results of the remaining analyses. The staff confirmed that 50 cycles was specified in the final safety analysis report and the applicant's use of at least 50 cycles of OBE transients is acceptable. The applicant committed (Regulatory Commitment No. 1) to perform the reanalysis for the applicable NUREG/CR-6260 locations and to submit the results to the staff, along with a summary of how the analysis was performed, no later than October 15, 2008. The applicant also committed to the use of a minimum of 50 cycles of the OBE transient and if needed, will report the use of less than 50 cycles along with the results. However, the staff noted that using less than 50 cycles of the OBE transient would not be consistent with the CLB. The applicant amended the LRA by letter dated October 2, 2008, and clarified that for the design fatigue analysis for the NUREG/CR-6260 locations utilized a minimum of 50 cycles of OBE (five events with ten cycles

4-51

each). The staff confirmed that the applicant amended its response to RAI 4.3-11 and has withdrawn Regulatory Commitment No. 1, as described in SER Section 4.3.3.3.2.

Based on its review, the staff finds that the applicant's response to RAI 4.3-11 is acceptable because the applicant has provided the requested information regarding the number of OBE transients that will be incorporated in the environmental assisted fatigue analyses (a minimum of 50 cycles of OBE, which is consistent with the CLB) and has committed to perform the reanalysis and has provided the results and method of analysis to the staff for the applicable NUREG/CR-6260 locations by letter dated October 2, 2008. Therefore, the staff's concern described in RAI 4.3-11 is resolved.

During the audit, the staff noted that the basis document WCAP-16173-P, Table 6-1, "Beaver Valley Units 1 and 2 Design Basis Transient Evaluation for License Renewal," March 2004, including Errata dated August 11, 2004 shows that the design cycles of OBE is 50 for several NSSS components of Unit 1, including the RV and pressurizer. The staff noted that annotation (d) of the LRA Table 4.3-2 states that the number of the design cycles for the OBE is 400 cycles for NSSS equipment and 50 cycles for the piping. In RAI 4.3-14, dated May 28, 2008, the staff requested that the applicant explain the discrepancy between LRA Table 4.3-2 and WCAP-16173-P, Table 6-1 and how the design cycles for the OBE will be considered in the CUF evaluation.

In its response to RAI 4.3-14, dated July 11, 2008, the applicant stated that the design cycles for the OBE listed in LRA Table 4.3-2 were taken from the UFSAR Table 4.1-10, Revision 24, for Unit 1. The applicant noted that WCAP-16173-P, Table 6-1 shows that the Unit 1 RV, pressurizer and steam generators were designed for 50 cycles of the OBE. The staff noted that the original steam generators were designed to 50 cycles of the OBE; however, the applicant has confirmed that the replacement steam generators have been designed for 400 cycles of the OBE. The applicant further noted that the information provided for the RV and pressurizer has been confirmed, and each is designed for 50 cycles of the OBE; and, the LRA has been amended to reflect this change. The staff noted that the applicant will address the error in UFSAR Table 4.1-10 for the Unit 1 under its corrective action program, which is subject to the 10 CFR 50.59 process. As described in the staff's evaluation of RAI 4.3-11, the applicant committed (Regulatory Commitment No. 1) to use a minimum of 50 cycles of the OBE when performing the CUF analyses for the NUREG/CR-6260 locations and to provide the results to the staff, along with a summary of how the analyses was performed, no later than October 15, 2008.

Based on its review, the staff finds the applicant's response to RAI 4.3-14 acceptable because the applicant will correct the discrepancy in the UFSAR for Unit 1 as part of its corrective action program, subject to a 10 CFR 50.59 review and has committed to perform the reanalysis for the applicable NUREG/CR-6260 locations and to provide to the staff, the results and method of analysis. Therefore, the staff's concern described in RAI 4.3-14 is resolved.

4.3.4.3 UFSAR Supplement

The applicant provided a UFSAR supplement summary description of its TLAA evaluation of the NSSS transient cycle projection for 60-year operation in LRA Sections A.2.3 and A.3.3. On the basis of its review of the UFSAR supplement, the staff concludes that the summary description of the applicant's actions to address NSSS transient cycle projection for 60-year operation is adequate.

4.3.4.4 Conclusion

Based on its review, as discussed above, the staff concludes that the applicant has demonstrated, pursuant to 10 CFR 54.21(c)(1)(i), that the analyses remain valid for the period of extended operation and, pursuant to 10 CFR 54.21(c)(1)(ii), that the analyses have been projected to the end of the period of extended operation. The staff further concludes, pursuant to 10 CFR 54.21(c)(1)(iii), that the effects of aging on the intended function(s) will be adequately managed for the period of extended operation. The staff also concludes that the UFSAR supplement contains an appropriate summary description of the TLAA evaluation, as required by 10 CFR 54.21(d).

4.4 Environmental Qualification of Electric Equipment

The 10 CFR 50.49 EQ program is a TLAA for purposes of license renewal. The TLAA of the EQ of electrical components includes all long-lived passive and active, electrical and instrumentation and control (I&C) components that are important to safety and are located in a harsh environment. The harsh environments of the plant are those areas subject to environmental effects by loss-of-coolant accidents (LOCAs) or high-energy line breaks (HELBs). EQ equipment comprises safety-related and Q-list equipment; nonsafety-related equipment, the failure of which could prevent satisfactory accomplishment of any safety-related function; and, necessary post-accident monitoring equipment.

As required by 10 CFR 54.21(c)(1), the applicant must provide a list of EQ TLAAs in the LRA. The applicant shall demonstrate that for each type of EQ equipment, one of the following is true: (1) the analyses remain valid for the period of extended operation, (2) the analyses have been projected to the end of the period of extended operation, or (3) the effects of aging on the intended function(s) will be adequately managed for the period of extended operation.

4.4.1 Summary of Technical Information in the Application

In LRA Section 4.4, the applicant summarized its evaluation of EQ of electrical equipment for the period of extended operation. The applicant's existing Environmental Qualification (EQ) of Electric Components Program manages component thermal, radiation and cyclical aging, as applicable, through the use of aging evaluations that are based on 10 CFR 50.49(f) qualification methods. As required by 10 CFR 50.49, EQ components not qualified for the current license term are to be refurbished, replaced, or have their qualification extended prior to reaching the aging limits established in the evaluation.

The Environmental Qualification (EQ) of Electric Components Program ensures that these EQ components are maintained in accordance with their qualification bases. Aging evaluations for EQ components that specify a qualification of at least 40 years are TLAAs for license renewal.

4.4.2 Staff Evaluation

The staff reviewed LRA Section 4.4 to verify, pursuant to 10 CFR 54.21(c)(1)(iii), that the effects of aging on the intended function(s) will be adequately managed for the period of extended operation.

The staff reviewed LRA Section 4.4 and plant basis documents to determine whether the applicant provided adequate information to meet the requirements of 10 CFR 54.21(c)(1). For the electrical equipments identified in the EQ master list, the applicant used 10 CFR 54.21(c)(1)(iii) in its TLAA evaluation to demonstrate that the aging effects of EQ equipment will be adequately managed during the period of extended operation. The staff reviewed the EQ program to determine whether it will assure that the electrical and I&C components covered under this program will continue to perform their intended functions, consistent with the CLB, for the period of extended operation. The staff's evaluation of the components qualification focused on how the EQ program manages the aging effects to meet the requirements of 10 CFR 50.49.

The staff conducted an audit of the information provided in LRA Section B 2.14 and the program basis documents. Based on its audit, the staff finds that the EQ program, which the applicant claimed to be consistent with GALL AMP X.E1, "Environment Qualification of Electrical Components," is consistent with EQ program in the GALL report. Therefore, the staff finds that the EQ program is capable of programmatically managing the qualified life of components within the scope of the program for license renewal. The continued implementation of the EQ program provides reasonable assurance that the aging effects will be managed and that components within the scope of the EQ program will continue to perform their intended functions for the period of extended operation.

4.4.3 UFSAR Supplement

The applicant provided a UFSAR supplement summary description of its TLAA evaluation of EQ of electrical equipment in LRA Section A.1.14. The UFSAR supplement is inconsistent with those in SRP-LR Table 4.4.2 in that it does not contain reanalysis attributes. Reanalysis addresses attributes of analytical methods, data collection and reduction methods, underlying assumptions, acceptance criteria, corrective actions if acceptance criteria are not met, and the period of time prior to the end of qualified life when the reanalysis will be completed. In RAI B.2.14-1, dated May 15, 2008, the staff requested that the applicant provide the important attributes of reanalysis of an aging evaluation in the UFSAR and the time when the reanalysis will be completed or provide a justification why it is not necessary to include these attributes in the UFSAR supplement.

In its response to RAI B.2.14-1, dated June 17, 2008, the applicant revised LRA Section A.1.14, "Environmental Qualification (EQ) of Electrical Components Program," to add additional details regarding the EQ component reanalysis attributes as detailed in GALL AMP X.E1, "Environmental Qualification (EQ) of Electric Components."

Based on its review of the UFSAR supplement, the staff concludes that the applicant's response to RAI B.2.14-1 and the summary description of its actions to address EQ of electrical equipment is adequate. Therefore, the staff's concern described in RAI B.2.14-1 is resolved.

4.4.4 Conclusion

Based on its review, the staff concludes that the applicant has demonstrated, pursuant to 10 CFR 54.21(c)(1)(iii), that for EQ of electrical equipment, the effects of aging on the intended function(s) will be adequately managed for the period of extended operation. The staff also concludes that the UFSAR supplement contains an appropriate summary description of the TLAA evaluation, as required by 10 CFR 54.21(d).

4.5 Concrete Containment Tendon Prestress

4.5.1 Summary of Technical Information in the Application

In LRA Section 4.5, a summary of evaluation of concrete containment tendon prestress for the period of extended operation is not applicable since Unit 1 and Unit 2 have no pre-stressed tendons in the containment building.

4.5.2 Staff Evaluation

The containment building has no prestressed tendons; therefore, the staff finds this TLAA is not required.

4.5.3 UFSAR Supplement

The staff concludes that no UFSAR supplement is required because the containment building has no pre-stressed tendons.

4.5.4 Conclusion

Based on its review, as discussed above, the staff concludes this TLAA is not required.

4.6 Containment Liner Plate, Metal Containment, and Penetrations Fatigue

4.6.1 Containment Liner Fatigue

In LRA Section 4.6.1, the applicant summarized the evaluation of containment liner fatigue for the period of extended operation. The function of the liner is to act as a gas tight membrane and no credit is taken for the liner's ability to resist primary bursting stresses. The applicant stated in the LRA that cyclic loads considered in the design of the liners for Units 1 and 2 include: (a) differential pressure cycling due to plant normal operation, namely startup and shutdown; (b) thermal cycling due to plant normal operation, namely startup and shutdown; and (c) stresses due to seismic cycling.

4.6.1.1 BVPS 1 Containment Liner

In LRA Section 4.6.1.1, the applicant stated that the Unit 1 containment liner stress analysis determines a fatigue usage factor based on specific design cyclic loads in accordance with ASME Code Section III, 1968 Edition, Paragraph N-415.2. In UFSAR Table 5.2-13 for Unit 1, the applicant noted 150 cycles of loading due to the differential pressure between operating and atmospheric pressure for a 60-year span, 600 cycles of loading due to thermal expansion resulting for a 60-year span, and 150 cycles of OBE for a 60-year span. The design limit includes 1000 cycles for operating pressure cycles, 4000 cycles for operating temperature variations, and 20 cycles for design basis earthquake. In the LRA, the applicant stated that the design cycles of the Unit 1 Containment liner bound the anticipated pressure and temperature cycles expected through the period of extended operation. The applicant further stated that the expected stresses resulting from the 60-year anticipated OBE cycles were determined to be bounded by those limits due to the analyzed design-basis earthquake cycles. Therefore, the Unit 1 containment liner fatigue TLAA has been dispositioned pursuant to 10 CFR 54.21(c)(1)(ii).

4.6.1.1.1 Staff Evaluation

The staff reviewed LRA Section 4.6.1, pursuant to 10 CFR 54.21(c)(1)(ii), to verify that the analyses have been projected to the end of the period of extended operation.

The staff's review of the SRP-LR Section 4.6.1 evaluation of the containment liner plates, metal containment, and penetrations fatigue analysis found that the applicant's code of record requires a fatigue analysis for the liner, from mechanical loadings in addition to thermal and anchor motion. For this reason, the staff reviewed the containment liner fatigue evaluation for the period of extended operation as required by 10 CFR 54.21(c). During its review, the staff reviewed UFSAR Table 5.2-13 for Unit 1 and some related onsite basis documents and found that both projected CUFs for 60 years. For pressure variation due to normal operations and temperature variation due to normal operations, CUFs are projected at 0.15. During a conference call held on October 8, 2008, the staff requested that the applicant address how the expected stresses resulting from the 60-year anticipated OBE cycles were bounded by those due to the analyzed design-basis earthquake cycles, since the design-basis earthquake cycles does not bound the 60-year anticipated OBE cycles. In the letter dated November 5, 2008, the applicant stated that the fatigue analysis determined the stress due to the combination of the thermal, normal operation and design-basis earthquake loadings. In determining the CUF, that combination was then considered as 4000 cycles of fluctuation from the operation condition (including design-basis earthquake) to the zero-stress state. The applicant further stated that the 60-year anticipated occurrence of 150 pressure cycles, 600 temperature cycles and 150 OBE cycles are bounded by the 4000 analyzed cycles. Therefore, the applicant concluded that no revision to the Unit 1 containment liner stress analysis was required and amended the LRA to change the TLAA disposition from 10 CFR 54.21(c)(1)(ii) to 10 CFR 54.21(c)(1)(i). The applicant also amended LRA Sections 4.6.1.1 and A.2.5.1 to reflect the change.

The staff reviewed supplement information and LRA amendment (Amendment No. 30) and finds that the applicant's assumption of 4000 combined cycles for normal operating, thermal, and design-basis earthquake loadings is conservative. The staff also confirms from the UFSAR for Unit 1 that expected stresses result from a combination of normal operating, thermal, and design-basis earthquake loadings. The staff determines that the 60-year anticipated occurrences of pressure cycles, temperature cycles and OBE cycles are bounded by the 4000 analyzed cycles and the projected CUF for 60 years is 0.225.

The staff reviewed the information presented in LRA Section 4.6.1.1 and finds that the applicant's containment liner stress analyses for Unit 1 follows the guidance of SRP-LR

Section 4.6.1. Therefore, the staff concludes that the existing analyses of fatigue for the Unit 1 containment liner will remain valid for the period of extended operation, in accordance with 10 CFR 54.21(c)(1)(i).

4.6.1.2 BVPS 2 Containment Liner

In LRA Section 4.6.1.2, the applicant stated that as a design guideline, the Unit 2 containment liner was designed in accordance with the ASME Code Section III, 1971 Edition, using stress limits and fatigue criteria based on the rules for ASME Code Classes MC and 1. The applicant further stated that a detailed analysis for fatigue is not required, if six specific requirements are met as defined in ASME Code Section III, NB-3222.4(d). In UFSAR Table 3.8-9 for Unit 2, the applicant has indicated 150 cycles of loading due to the differential pressure between operating and atmospheric pressure for a 60-year span, 600 cycles of loading due to thermal expansion resulting for a 60-year span, and 150 cycles of OBE for a 60-year span.

4.6.1.2.1 Staff Evaluation

The staff reviewed LRA Section 4.6.1.2, pursuant to 10 CFR 54.21(c)(1)(ii), to verify that the analyses has been projected to the end of the period of extended operation.

The staff's review of the SRP-LR Section 4.6.1 evaluation of the containment liner plates, metal containment, and penetrations fatigue analysis found that the applicant's code of record requires a fatigue analysis for the liner, from mechanical loadings in addition to thermal and anchor motion. For this reason, the staff reviewed the containment liner fatigue evaluation for the period of extended operation as required by 10 CFR 54.21(c). During the review, the staff found that the stress limits and fatigue criteria of the Unit 2 containment liner follow the design guidelines in accordance with ASME Code Section III, 1971 Edition. The staff confirmed that the ASME Code Section III does not require a detailed fatigue analysis, if six specific requirements are met as defined in Subsection NB-3222.4(d). The staff also reviewed the Unit 2 UFSAR Table 3.8-9 and found that the design limit includes 1000 cycles for operating pressure cycles, 4000 cycles for operating temperature variations, and 20 cycles for safe shutdown earthquake. The staff reviewed the applicant's re-evaluation of the six fatigue exemption requirements, utilizing anticipated 60-year stress cycles, and determined that extended operation continues to satisfy the requirement for exemption from a detail fatigue analysis for cyclic operation. Therefore, the staff confirms that the design load cycles will not be reached by the anticipated 60-year load cycles.

The staff reviewed LRA Section 4.6.1.2 and the relevant references cited in the TLAA and finds that the applicant's containment liner stress analyses for Unit 2 follows the guidance of SRP-LR Section 4.6.1. Therefore, the staff concludes that the analyses of fatigue for the Unit 2 containment liner have been projected to the end of the period of extended operation in accordance with 10 CFR 54.21(c)(1)(ii).

4.6.1.3 UFSAR Supplement

The applicant provided a UFSAR supplement summary description of its TLAA evaluation of the Unit 1 containment liner fatigue in LRA Section A.2.5.1 (Amendment No. 30). In LRA Section A.3.5.1, the applicant also provided a UFSAR supplement summary description of its TLAA evaluation of the Unit 2 containment liner fatigue. Based on its review of the UFSAR supplements, the staff concludes that the summary description of the applicant's actions to address the containment liner fatigue is adequate because the applicant's summary descriptions conform to the staff's guidance in SRP-LR Section 4.6, Table 4.6-1.

4.6.1.4 Conclusion

Based on its review, the staff concludes that the applicant has demonstrated, pursuant to 10 CFR 54.21(c)(1)(i) and 10 CFR 54.21(c)(1)(ii), that for the containment liner fatigue TLAA, the analyses of Unit 1 containment liner remain valid through the period of extended operation, and the analyses of Unit 2 containment liner has been projected to the end of the period of extended operation. The staff also concludes that the FSAR supplement contains an appropriate summary description of the TLAA evaluation, as required by 10 CFR 54.21(d).

4.6.2 Containment Liner Corrosion Allowance

4.6.2.1 Summary of Technical Information in the Application

In LRA Section 4.6.2, the applicant summarized the evaluation of containment liner corrosion allowance for the period of extended operation. The containment buildings for Units 1 and 2 have a continuously welded carbon steel liner which acts as a leak tight membrane. All welded seams were originally covered with continuously welded leak test channels which were installed to facilitate leak testing of welds during liner erection. Some vent plugs in the containment floor liner test channels for Units 2 and 1 were found missing in 1990 and 1991, respectively. The missing test channel vent plugs allowed moisture and condensation inside the test channels, leading to minor corrosion of the liner. In the LRA, the applicant stated that the test channels were evaluated to demonstrate that the corrosion rates inside the test channels would not result in the liner failing to meet its minimum required thickness based on a 40-year corrosion period. The applicant stated that these corrosion rate analyses meet the requirements of 10 CFR 54.3 as a TLAA and must be evaluated for the period of extended operation.

The applicant further stated that the corrosion allowance for the containment floor liners has a fabrication thickness of 0.25 inches and a minimum required thickness of 0.125 inches (both units). Thus, the corrosion allowance is 0.125 inches (125 mils). The total estimated penetration due to corrosion of the inerted channel was estimated at 69.2 mils and 82.7 mils for 40 years of plan operation and 3 years of pre-operational exposure for Units 1 and 2, respectively. The projected 60-year corrosion penetration depths yield 77.0 mils and 90.5 mils for Units 1 and 2, respectively.

4.6.2.2 Staff Evaluation

The staff reviewed LRA Section 4.6.2, pursuant to 10 CFR 54.21(c)(1)(ii), to verify that the analyses have been projected to the end of the period of extended operation.

The SRP-LR Section 4.1, Table 4.1-2, states that metal corrosion allowance is a potential TLAA. For this reason, the staff reviewed the containment floor liner corrosion evaluation for the period of extended operation as required by 10 CFR 54.21(c). The staff reviewed the related onsite basis documents and found that the applicant had calculated in one of the documents a different corrosion allowance value of 88 mils, instead of 125 mils. In RAI 4.6.2-1, dated May 8, 2008, the staff requested that the applicant explain the discrepancy of the corrosion allowance for the liner floor plate.

In its response to RAI 4.6.2-1, dated June 16, 2008, the applicant stated that the corrosion allowance of 88 mils was based on corrosion rate information published by the General Electric Corporation. The applicant explained that the basis document reviewed by the staff was a report prepared in March 1991. This and earlier reports used the 88 mils corrosion allowance in the context that there is sufficient margin in the containment liner thickness to easily accommodate a corrosion of 88 mils. The applicant further stated that liner floor plate minimum wall thickness of 125 mils was established by design analysis calculations, which provide a corrosion allowance of 125 mils out of a 0.25 inch plate and is the CLB.

Based on its review, the staff finds the applicant's response to RAI 4.6.2-1 acceptable because the applicant has clarified the discrepancy of the corrosion allowance for the liner floor plate. The staff confirms that the applicant's projected 60-year corrosion penetration depths yield 77.0 mils or 62% of the corrosion allowance, and 90.5 mils or 72% of corrosion allowance for Units 1 and 2, respectively. The staff determines that the applicant's containment liner corrosion analyses for Units 1 and 2 have been projected to the end of the period of extended operation, in accordance with 10 CFR 54.21(c)(1)(ii). Therefore, the staff's concern described in RAI 4.6.2-1 is resolved.

4.6.2.3 UFSAR Supplement

The applicant provided UFSAR supplement summary descriptions of the TLAA evaluation of the containment linear corrosion allowance in LRA Sections A.2.5.2 and A.3.5.2. Based on its review of the UFSAR supplement, the staff concludes that the summary description of the applicant's actions to address the containment linear corrosion allowance is adequate because the applicant's summary description conforms to the guidance found in SRP-LR Section 4.6, Table 4.6-1.

4.6.2.4 Conclusion

Based on its review, the staff concludes that the applicant has demonstrated, pursuant to 10 CFR 54.21(c)(1)(ii), that for the containment linear corrosion allowance TLAA, the analyses have been projected to the end of the period of extended operation. The staff also concludes that the FSAR supplement contains an appropriate summary description of the TLAA evaluation, as required by 10 CFR 54.21(d).

4.6.3 Containment Liner Penetration Fatigue

4.6.3.1. BVPS 1 Containment Liner Penetration Fatigue

4.6.3.1.1 Staff Evaluation

In LRA Section 4.6.3.1, the applicant summarized the Unit 1 containment liner penetration fatigue as follows:

> (1) Cold penetrations have the process pipe welded to a plate flange which is anchored to the containment concrete wall such that loads can be

transferred from the piping to the reinforced concrete. Hot penetrations (> 180°F) are designed with a sleeve and liner such that water-cooled cooling units and appropriate insulation can be located inside the annulus to maintain the concrete temperature within allowable levels. UFSAR Section 5.2.4.8, indicated that the evaluation of the penetration discontinuities was done using a computer program entitled SHELL-1, which analyzes axisymmetric thin shells of revolution under unsymmetrical loading. The applicant evaluated the temperature distribution at discontinuity areas exposed to operating conditions using finite difference or finite elements techniques. While ASME Code Section III was used as a guide in the selection of design stresses used in the analysis of these penetrations, no specific fatigue analysis was completed for the Unit 1 piping penetrations. Therefore, no TLAA is associated with the Unit 1 piping penetrations.

(2) The equipment hatch and integral emergency airlock are designed and analyzed in accordance with ASME Code Section III, Division 1, Subsection NE (Class MC). The applicant completed a fatigue exemption the equipment hatch in accordance with Subsection NB-3222(d). This exemption was based on assumed cycles for a 40-year life, namely, 10 pressurization events due to LOCA and 80 cycles of startup and shutdown. The applicant stated that it is highly unlikely that Unit 1 will reach 10 pressurization events due to LOCA during 60-years of operation. The assumption of 80 cycles of startup and shutdown is not bounding for 60 years of operation. The applicant performed a reanalysis using 240 startup and shutdown cycles that bounds the number of projected cycles for the period of extended operation. Therefore, the equipment hatch fatigue TLAA has been dispositioned in accordance with 10 CFR 54.21(c)(1)(ii).

(3) The applicant analyzed the personnel air lock pursuant to ASME Code Section III, Class B. However, no fatigue analysis was completed for this air lock. Therefore, no TLAA is associated with the personnel air lock.

(4) The applicant analyzed the fuel transfer tube pipe pursuant to ASME Section III, Division 1, Subsection NC. The analysis for the fuel transfer tube pipe uses a stress range reduction factor of 1.0 (<7,000 cycles). However, since the fuel transfer tube pipe experiences operational cycles only during refueling, the fuel transfer tube pipe essentially experiences no thermal cycles. The applicant concluded that the existing fuel transfer tube pipe stress analysis remains valid through the period of extended operation. Therefore, the fuel transfer tube pipe fatigue TLAA has been dispositioned in accordance with 10 CFR 54.21(c)(1)(i).

(5) The applicant also analyzed the fuel transfer tube bellows pursuant to ASME Section III, Division 1, Subsection NC. The applicant stated that the bellows stress analysis was acceptable on the basis that the bellows

experienced displacements due to a design-basis earthquake (DBE). The analysis assumed 600 design. The applicant further stated that the occurrence of this number of DBE cycles is highly unlikely during the period of extended operation. The applicant concluded that the fuel transfer tube bellows stress analysis remains valid through the period of extended operation. Therefore, the fuel transfer tube bellows fatigue TLAAs have been dispositioned in accordance with 10 CFR 54.21(c)(1)(i).

4.6.3.1.2 UFSAR Supplement

During the audit and review, the staff reviewed LRA Section 4.6.3.1, pursuant to 10 CFR 54.21 (c)(1)(i), to verify that the analyses will remain valid for the period of extended operation, and pursuant to 10 CFR 54.21(c)(1)(ii), to verify that the analyses have been projected to the end of the period of extended operation.

The staff reviewed SRP-LR Section 4.6.1 for the containment liner penetration fatigue analysis and found that the applicant's code of record requires a fatigue analysis for the containment liner penetration from mechanical loadings as well as thermal and anchor motion. For this reason and to comply with 10 CFR 54.21(c), the staff reviewed the containment liner penetration fatigue evaluation for the period of extended operation and found the following:

(1)   The staff reviewed the Unit 1 UFSAR Section 5.2.4.8 and determined that the applicant had performed an evaluation using a finite difference or finite elements techniques. The staff found that the applicant had not completed a specific fatigue analysis for the Unit 1 piping penetrations; and, therefore, no TLAA is associated with the Unit 1 piping penetrations. The staff confirmed with the applicant that no TLAA was associated with the Unit 1 piping penetrations and personnel air lock.

(2)   The staff reviewed the plant-specific analysis documents for the equipment hatch and found that the applicant performed a reanalysis using 240 startup and shutdown cycles for a projected CUF for 60 years of 0.33. Therefore, the analyses of fatigue for the Unit 1 containment liner have been projected to the end of the period of extended operation in accordance with 10 CFR 54.21(c)(1)(ii).

(3)   The staff also found the applicant did not complete a fatigue analysis for the air lock. Therefore, no TLAA is associated with the Unit 1 personnel air lock.

(4)   Since the Unit 1 fuel transfer tube pipe experiences operational cycles only during refueling, essentially no thermal cycles are experienced. However, the applicant's analysis uses a stress range reduction factor of 1.0 (< 7,000 cycles) pursuant to ASME Code Section III, Division 1, Subsection NC; therefore, the existing fuel transfer tube pipe stress analysis remains valid through the period of extended operation and, the Unit 1 fuel transfer tube pipe fatigue TLAA has been dispositioned in accordance with 10 CFR 54.21(c)(1)(i).

(5)   The staff also noted that the Unit 1 fuel transfer tube bellows were analyzed pursuant to ASME Section III, Division 1, Subsection NC to determine acceptability based on the bellows experiencing displacements due to a DBE. The assumed design cycles were 600. The occurrence of this number of DBE cycles is highly unlikely during the period of extended operation. Therefore, the Unit 1 fuel transfer tube bellows fatigue TLAA has been dispositioned in accordance with 10 CFR 54.21(c)(1)(i).

The staff concludes that the information presented in LRA Section 4.6.3.1 and the relevant references cited in the TLAA are acceptable and finds that they meet the requirements of SRP-LR Section 4.6.1. Therefore, the fatigue analyses of fatigue for the Unit 1 containment liner penetrations have been projected to the end of the period of extended operation in accordance with 10 CFR 54.21(c)(1)(i) and 10 CFR 54.21(c)(1)(ii).

4.6.3.1.3 Conclusion

Based on its review, the staff concludes that the applicant has demonstrated, pursuant to 10 CFR 54.21(c)(1)(i) and 10 CFR 54.21(c)(1)(ii), that for the containment liner penetration fatigue TLAA, the analyses correspondingly either remain valid through the period of extended operation or has been projected to the end of the period of extended operation. The staff also concludes that the FSAR supplement contains an appropriate summary description of the TLAA evaluation, as required by 10 CFR 54.21(d).

4.6.3.2 BVPS 1 Containment Penetration Bellows

In LRA Section 4.6.3.2, the applicant summarized the Unit 1 containment penetration bellows and stated that the Unit 1 containment penetration bellows are part of the system evaluation boundary of the Unit 1 river water system. The piping and in-line components of the Unit 1 river water system are designed and analyzed in accordance with the ANSI B31.1 standard, which specifies evaluation of cyclic secondary stresses by applying stress range reduction factors against the allowable stress range. The assumed design limit is 7,000 thermal cycles. The staff noted that the Unit 1 recirculation system fatigue analyses remain valid for the period of extended operation in accordance with 10 CFR 54.21(c)(1)(i).

4.6.3.2.1 Staff Evaluation

The staff reviewed LRA Section 4.6.3.2, pursuant to 10 CFR 54.21 (c)(1)(i), to verify that the analyses will remain valid for the period of extended operation.

The staff's review of the SRP-LR Section 4.6.1 evaluation of the containment liner penetration bellows analysis found that the applicant's code of record requires a fatigue analysis for the liner from mechanical loadings as well as thermal and anchor motion. For this reason, the staff reviewed the containment liner fatigue evaluation for the period of extended operation, as required by 10 CFR 54.21(c). During the review, the staff confirmed that the stress analysis of the Unit 1 containment penetration bellows follows the ANSI B31.1, 1967 Addition standard. The staff noted that the Unit 1 recirculation spray system normally is in standby operation and including any periodic tests, will experience significantly less than the 7,000 full temperature cycle limits for the period of extended operation.

Therefore, the Unit 1 containment liner penetration bellows TLAA analyses remain valid for the period of extended operation, in accordance with 10 CFR 54.21(c)(1)(i).

4.6.3.2.2 UFSAR Supplement

The applicant provided a UFSAR supplement summary description of its TLAA evaluation of the Unit 1 containment liner penetration fatigue in LRA Section A.2.5.3. Based on its review of the UFSAR supplement, the staff concludes that the summary description of the applicant's actions to address the containment liner penetration fatigue is adequate because the applicant's summary description conforms to the guidance found in SRP-LR Section 4.6, Table 4.6-1.

4.6.3.2.3 Conclusion

Based on its review, the staff concludes that the applicant has demonstrated, pursuant to 10 CFR 54.21(c)(1)(i) that for the containment liner penetration bellows TLAA, the analyses have been projected to the end of the period of extended operation. The staff also concludes that the FSAR supplement contains an appropriate summary description of the TLAA evaluation, as required by 10 CFR 54.21(d).

4.6.3.3 BVPS 2 Containment Liner Penetration Fatigue

In LRA Section 4.6.3.3, the applicant summarized the Unit 2 Containment Liner Penetration Fatigue as follows:

> (1) The applicant designed and analyzed the Unit 2 process piping penetrations in accordance with the ASME Code Section III, Division 1, 1971 Edition through the 1972 Winter Addenda, Subsection NC (Class 2), which complies with the process piping system requirements of which these penetrations are a part. The applicant further analyzed the penetrations in accordance with the more rigorous Subsection NE (Class MC) requirements. ASME Code Section III, Division 1, Class 2 requirements include a stress range reduction factor which accounts for an assumed number of thermal cycles. In addition, the applicant performed a fatigue exemption of the Class MC portion of the stress analysis in accordance with ASME Code Section III, Subsection NE-3322(d) and evaluated the validity of this assumption for 60 years of plant operation. The results of the evaluation indicate that the thermal cycle assumption is valid and bounding for 60 years of operation. Therefore, the applicant determined that the piping penetration fatigue analyses remain valid for the period of extended operation, in accordance with 10 CFR 54.21(c)(1)(i).

> (2) The applicant designed and analyzed the Unit 2 containment equipment hatch and integral emergency airlock in accordance with ASME Code Section III, Division 1, Subsection NE (Class MC), 1971

Edition through 1972 Winter Addenda. The applicant stated that a reanalysis, using projected 60-year startup and shutdown cycles, was performed for the period of extended operation to confirm the fatigue exemption described in ASME Code, Subsection NB-3222(d), in accordance with 10 CFR 54.21(c)(1)(ii).

(3)   The applicant analyzed the personnel air lock in accordance with ASME Code Section III, Division 1, Subsection NE (Class MC) but did not complete a fatigue analysis for this air lock. Therefore, no TLAA is associated with the Unit 2 personnel air lock.

(4)   The applicant designed and analyzed the Unit 2 fuel transfer tube pipe in accordance with ASME Code Section III, Class 2 (Subsection NC). The applicant stated that the design cycles of the Unit 2 fuel transfer tube pipe, relative to the stress range reduction factor of 1.0 (<7,000 cycles), bound the anticipated 60-year Unit 2 refueling cycles expected through the period of operation. The applicant determined that the existing fuel transfer tube pipe stress analysis remains valid through the period of extended operation, and dispositioned the Unit 2 fuel transfer tube pipe fatigue TLAA in accordance with 10 CFR 54.21(c)(1)(i).

(5)   The applicant designed and analyzed the Unit 2 fuel transfer tube bellows in accordance with ASME Section III, Class MC. The applicant's bellows stress analysis determined acceptability based on the bellows experiencing displacements due to a design basis earthquake. The applicant assumed 600 design cycles in its analysis. The applicant determined that the existing fuel transfer tube bellows stress report remains valid through the period of extended operation, and dispositioned the Unit 2 fuel transfer tube bellows fatigue TLAAs in accordance with 10 CFR 54.21(c)(1)(i).

4.6.3.3.1 Staff Evaluation

During its review, the staff reviewed LRA Section 4.6.3.3, pursuant to 10 CFR 54.21 (c)(1)(i), to verify that the analyses will remain valid for the period of extended operation, and pursuant to 10 CFR 54.21 (c)(1)(ii), to verify that the analyses have been projected to the end of the period of extended operation.

The staff review of the SRP-LR Section 4.6.1 evaluation of the containment liner penetration fatigue analysis found that the applicant's code of record requires a fatigue analysis for the liner from mechanical loadings as well as thermal and anchor motion. For this reason, the staff reviewed the containment liner fatigue evaluation for the period of extended operation, as required by 10 CFR 54.21(c) and determined the following:

(1)   The staff noted that as a design guideline, the stress limits and fatigue criteria of the Unit 2 process piping penetrations follow the ASME Code Section III, Division 1, Subsection NC (Class 2). The staff verified that

the applicant has further analyzed the penetrations in accordance with the more rigorous Subsection NE (Class MC) requirements which allow an exemption of a detailed fatigue analysis, if specific requirements are met as defined in ASME Code Section III, NB-3222(d). The staff reviewed the applicant's existing evaluation of the fatigue exemption requirements based on the LRA and plant-specific analysis documents, and confirmed that the applicant's assumed full-temperature design cycles bound the anticipated 60-year significant thermal cycles of Unit 2 process piping penetrations, including: (a) unsleeved penetrations; (b) sleeved piping penetrations; and (c) multiple piping penetrations. Therefore, the staff determines that the applicant's existing Unit 2 process piping penetration fatigue analyses will remain valid for the period of extended operation, in accordance with 10 CFR 54.21(c)(1)(i).

(2) The staff confirmed that the applicant has designed and analyzed the Unit 2 containment equipment hatch and integral emergency airlock in accordance with ASME Code Section III, Division 1, Subsection NE (Class MC), which exempts a detailed analysis for fatigue, if specific requirements are met in accordance with ASME Code Section III, NB-3222(d). The staff noted that the applicant's assumption of 10 pressurization events due LOCA for 40-year life is bounding for 60 years of operation. The staff also reviewed the applicant's plant-specific analysis documents for the equipment hatch and found that a reanalysis was performed using 240 startup and shutdown cycles that bounds the number of projected cycles for the period of extended operation, instead of the its assumed 80 cycles for a 40-year life. Therefore, the staff determines that the applicant's the analyses of fatigue for the Unit 2 containment equipment hatch have been projected to the end of the period of extended operation, in accordance with 10 CFR 54.21(c)(1)(ii).

(3) The staff also confirmed that no fatigue analysis was required for the air lock. Therefore, the staff determines that no TLAA is associated with the Unit 2 personnel air lock.

(4) The staff agreed with the applicant that the design cycles of the Unit 2 fuel transfer tube pipe in terms of stress range reduction factor of 1.0 (<7,000 cycles) bound the anticipated 60-year Unit 2 refueling cycles expected through the period of operation, because the Unit 2 fuel transfer tube pipe experiences operational cycles only during refueling, and essentially experiences no thermal cycles. Therefore, the staff determines that the applicant's existing fuel transfer tube pipe stress analysis remains valid through the period of extended operation, in accordance with 10 CFR 54.21(c)(1)(i).

(5) The staff noted that as a design guideline, the applicant's stress analysis of the Unit 2 fuel transfer tube bellows follows ASME Code Section III, Class MC requirements to determine acceptability based on the bellows experiencing displacements due to a design-basis earthquake. The applicant assumed 600 cycles for design-basis earthquake. The staff confirms that the assumed number of design-basis earthquake cycles is

unlikely to occur during the period of extended operation. For that reason, the staff agrees that the assumed 600 design-basis earthquake bounds the anticipated 60-year Unit 2 OBE cycles expected through the period of operation. Therefore, the staff determines that the applicant's existing fuel transfer tube bellows stress report remains valid through the period of extended operation in accordance with 10 CFR 54.21(c)(1)(i).

The staff concludes that the information presented in LRA Section 4.6.3.3 and the relevant references cited in the TLAA follows the guidance of SRP-LR Section 4.6.1. Therefore, the applicant's analyses of fatigue for the Unit 2 containment liner penetrations either remain valid through the period of extended operation or have been projected to the end of the period of extended operation, in accordance with 10 CFR 54.21(c)(1)(i) and 10 CFR 54.21(c)(1)(ii), respectively.

4.6.3.3.2 UFSAR Supplement

The applicant provided a UFSAR supplement summary description of its TLAA evaluation of the Unit 2 containment linear penetration fatigue in LRA Section A.3.5.3. Based on its review of the UFSAR supplement, the staff concludes that the summary description of the applicant's actions to address the containment linear penetration fatigue is adequate because the applicant's summary description conforms to the guidance found in SRP-LR Section 4.6, Table 4.6-1.

4.6.3.3.3 Conclusion

Based on its review, the staff concludes that the applicant has demonstrated, pursuant to 10 CFR 54.21 (c)(1)(i) and 10 CFR 54.21(c)(1)(ii), that for the containment linear penetration fatigue TLAA, the analyses correspondingly either remain valid through the period of extended operation or have been projected to the end of the period of extended operation. The staff also concludes that the FSAR supplement contains an appropriate summary description of the TLAA evaluation, as required by 10 CFR 54.21(d).

4.7 Other Time-Limited Aging Analyses

In LRA Section 4.7, the applicant summarized its evaluations of the following plant-specific TLAAs:

- piping subsurface indications (Unit 1 only)
- reactor vessel underclad cracking (Unit 1 only)
- leak-before-break
- high-energy line break postulation
- settlement of structures (Unit 2 only)
- crane load cycles

4.7.1 Piping Subsurface Indications (Unit 1 Only)

4.7.1.1 Summary of Technical Information in the Application

During the inservice inspection (ISI) for Unit 1 at RFO 11 (March to May 1996), the applicant detected an indication in a weld which joined an elbow and a Section of straight pipe on the

RCS Loop C cold leg that exceeded the acceptance criteria of ASME Code, Section XI, Subsection IWB-3500. The applicant noted that the composition of the Section of straight pipe is Class 1 cast austenitic stainless steel (CASS).

The applicant performed a flaw evaluation pursuant to ASME Code, Section XI, Subsection IWB3600, with support from Appendix C; and, concluded that the postulated flaw met the Code requirements with significant margins of safety to the end of the service lifetime. By letter dated May 1, 1996, the staff approved the applicant's flaw evaluation. The staff further determined that this evaluation is a TLAA because the two parameters used in the evaluation, namely, thermal aging and fatigue transient cycles, are based on the service life of the piping (*i.e.*, time-dependent).

Thermal aging of CASS material continues until it reaches the saturation or fully-aged point. The limiting fracture toughness properties are those of the straight pipe, which has relatively high ferrite content. Therefore, the applicant used the fully-aged (saturated) fracture toughness properties of the straight pipe in its flaw evaluation. The applicant stated that because fully-aged stainless steel material properties were used, its flaw evaluation has no material property time-dependency requiring further TLAA evaluation for license renewal.

The applicant postulated an initial flaw and calculated the crack growth based on imposed loading transients. The transient cycles assumed in the flaw evaluation are conservative compared to the original design cycles. In LRA Table 4.3-2, the applicant showed the original design-basis transients, including RCS design cycles, along with the projected operational cycles that it anticipates will occur for 60 years of plant life. Based on projected operational cycles to 60 years, the applicant has determined that the design cycles are bounding for the period of extended operation.

The applicant stated that because the flaw growth evaluation remains valid for 60 years based on the 60-year projected operational cycles, the Metal Fatigue of Reactor Coolant Pressure Boundary Program will be used to validate the cycles assumed in the flaw evaluation.

4.7.1.2 Staff Evaluation

The staff reviewed LRA Section 4.7.1 to verify, pursuant to 10 CFR 54.21(c)(1)(i), that the analyses have been projected to the end of the period of extended operation and, pursuant to 10 CFR 54.21(c)(1)(iii), that the effects of aging on the intended function(s) will be adequately managed for the period of extended operation.

In LRA Section 4.7.1, the applicant stated that an indication was identified on the RCS loop C cold leg between an elbow and a Section of straight pipe. The staff determined that additional information was required to complete the its review.

In RAI 4.7.1-1, dated April 1, 2008, the staff requested that the applicant (a) explain the inspection history and results of the indication; (b) confirm the future inspection frequency of the indication; (c) clarify the exact location of the subject indication (*e.g.*, in the weld that joins the elbow and pipe, on the elbow, or on the pipe), and (d) explain the degradation mechanism of the indication.

In its response to RAI 4.7.1-1, dated June 2, 2008, the applicant:

- explained in part (a) that a flaw indication for weld DLW-LOOP3-7-S-02 was identified during a Unit 1 ISI examination performed in 1996. The applicant provided the dimensions for a flaw that would bound the indications found during the examination. The applicant indicated that since the flaw indication exceeded the ASME Code IWB-3500, an analysis was performed and approved by the staff to ensure that flaw met the applicable requirement with significant margins of safety to the end of service lifetime.

- confirmed in part (b) that a relief request was submitted to the staff. The staff substantially granted the applicant's relief request, which the applicant noted that the subject weld is not scheduled for future examinations.

- clarified in part (c) that the weld is the first circumferential weld after the RV nozzle-to-safe-end weld on the RCS C loop cold leg (the other end of the elbow is welded to the Reactor Vessel safe-end).

- explained in part (d) that the degradation is not caused by stress-corrosion cracking based on the conclusion reached in the flaw analysis for the subject weld.

The staff reviewed the applicant's responses and the accompanying references for the responses and confirms that the staff has approved both the evaluation of flaw indication in the reactor coolant system (RCS) cold leg pipe weld and the relief request for an alternative risked-informed ISI.

Based on its review, the staff finds the applicant's response to RAI 4.7.1-1 acceptable because the applicant has adequately (a) explained the inspection history and results of the indication; (b) confirmed the future inspection frequency of the indication; (c) clarified the exact location of the subject indication (e.g., in the weld that joins the elbow and pipe, on the elbow, or on the pipe), and (d) discussed the degradation mechanism of the indication. Therefore, the staff's concern described in RAI 4.7.1-1 is resolved.

In LRA Section 4.7.1, the applicant indicated that a flaw occurred between an elbow and a Section of straight pipe, which is made up Class 1 CASS piping. The staff was unclear as to the material specification and the indication characterization, and determined that additional information was required to complete the evaluation.

In RAI 4.7.1-2, dated April 1, 2008, the staff requested that the applicant (a) confirm that the elbow is made of CASS; (b) verify the material specification of the weld that joins the elbow and the pipe; (c) describe the indication size and characterization, and (d) justify the reliability and accuracy of the detection and characterization of the subject indication, since ultrasonic testing (UT) of CASS material cannot be performed to meet the requirements of ASME Code, Section XI, Appendix VIII.

In its response to RAI 4.7.1-2, dated June 2, 2008, the applicant:

- confirmed in part (a) that elbow was fabricated from Grade CF8M CASS.

- verified in part (b) that the weld filler material is TP 308 stainless steel. The applicant also noted that the weld was made using a tungsten inert gas weld process for the root passes.

4-68

- described in part (c) that ISI examinations revealed the presence of four inside diameter (ID) indications. In addition, the applicant stated that the four indications were considered bounded by a single composite flaw, for which the applicant also provided dimensions.

- justified in part (d) how each of the three examinations was performed. In addition, the applicant provided the reference for a SER in which the staff confirmed the presence of this flaw indication during an independent inspection on April 25, 1996.

The staff reviewed the applicant's response as well as the abovementioned SER. The staff notes that given the difficult conditions for accurate interpretation of the indication from UT, the applicant selected a flaw size that bounded all four indications. This approach, the staff notes, provided a conservative application of the flaw shape requirements of the ASME Code Section XI, Appendix C. In addition, the staff confirms that the abovementioned SER provided the dimension of the flaw based on staff inspection and concludes that it is comparable to that selected by the applicant.

Based on its review, the staff finds the applicant's response to RAI 4.7.1-2 acceptable because the applicant has adequately: (a) confirmed that the elbow is made of CASS; (b) verified the material specification of the weld that joins the elbow and the pipe; (c) described the indication size and characterization, and (d) justified the reliability and accuracy of the detection and characterization of the subject indication. Therefore, the staff's concern described in RAI 4.7.1-2 is resolved.

In LRA Section 4.7.1, the applicant indicated that the fully aged fracture toughness properties of the CASS straight pipe were used in the flaw evaluation. However, it was not clear to the staff whether the CASS piping has more limiting material properties compared to the subject weld. The staff determined that additional information was required to complete its review. In RAI 4.7.1-3, dated April 1,, 2008, the staff requested that the applicant confirm that the fracture toughness properties of the CASS piping are more limiting than the fracture toughness of the weld at the 60 years.

In its response to RAI 4.7.1-3, dated June 2, 2008, the applicant provided the fracture toughness values for the full service life of elbow, weld, and piping. The applicant noted that these values were used in the flaw analysis submitted to and approved by the staff. The staff confirms that the fracture toughness value for the piping is the most limiting and; therefore, appropriately used by the applicant in the flaw analysis.

Based on its review, the staff finds the applicant's response to RAI 4.7.1-3 acceptable because the applicant has confirmed that the fracture toughness properties of the CASS piping are more limiting than the fracture toughness of the weld at the 60 years. Therefore, the staff's concern described in RAI 4.7.1-3 is resolved.

In LRA Section 4.7.1, the applicant stated that the flaw evaluation includes the postulation of an initial flaw and the growth of that flaw based on imposed loading transients. However, the applicant did not include any information on the initial flaw size, the flaw growth rate, as well as loading transients. The staff was unable to conclude that flaw size will be within applicable limits at the end of the 60 years, and required additional information to complete its review.

In RAI 4.7.1-4, dated April 1, 2008, the staff requested that the applicant (a) explain the initial flaw size assumed in the analysis; (b) explain the flaw growth rate used and associated references; and (c) verify whether the operating cycles used in flaw evaluation performed in 1996 exceed the projected operational cycles at the end of 60 years.

In its response to RAI 4.7.1-4, dated June 2, 2008, the applicant:

- explained in part (a) that the initial flaw size assumed in the flaw analysis included flaw depth, length, thickness, depth parameter, and shape parameter.

- explained in part (b) that relevant data was used for austenitic stainless steels in a PWR water environment in an article from the Journal of Pressure Vessel Technology. In addition, the applicant stated that it used the appropriate crack growth rate equation. The staff notes that the above information was submitted by the applicant as part of flaw analysis. The staff subsequently approved the flaw analysis and documented its evaluation dated May 1, 1996. The staff verified in part (c) that the transients used in the flaw analysis are the design transients contained in the equipment specification. In addition, the applicant stated that Units 1 and 2 were designed to a slightly different set of transients than those used in the flaw analysis, and that these transient were considered to be conservative relative to the original design transients.

The staff reviewed the list of transients along with their cycle numbers considered in the flaw analysis and compared them with the Unit 1 original transients in LRA Table 4.3-2. The staff agreed with the applicant's claim that the flaw analysis transients were conservative relative to the original design transients. The staff notes that none of the original transients exceeded their design cycles at the end of 60 years, and concluded that the transients considered by the applicant in the flaw analysis will not exceed its design cycles.

Based on its review, the staff finds the applicant's response to RAI 4.7.1-4 acceptable because the applicant has (a) adequately explained the initial flaw size assumed in the analysis; (b) adequately explained the flaw growth rate used and associated references; and (c) verified that the operating cycles used in flaw evaluation performed in 1996 do not exceed the projected operational cycles at the end of 60 years. Therefore, the staff's concern described in RAI 4.7.1-4 is resolved.

In LRA Section 4.7.1, the applicant described a flaw indication identified on the RCS loop C cold leg between an elbow and a Section of straight pipe. The staff was unclear whether additional flaw indications existed at Units 1 and 2, and required additional information to complete its review.

In RAI 4.7.1-5, dated April 1,, 2008, the staff requested that the applicant: (a) identify all Class 1 components that contain indications or flaws that have remained in service at Units 1 and 2; (b) briefly explain the flaw evaluations performed for the affected components in accordance with the ASME Code, Section XI; (c) discuss how the indications or flaws were accepted and reference the appropriate ASME Code requirements; (d) provide indication and/or flaw characterization; and (e) discuss the analyses performed to accept the degraded components for the extended period of operation.

In its response to RAI 4.7.1-5, dated June 2, 2008, the applicant stated that LRA Section 4.7.1 only documents reportable flaws left-in-service with acceptance provided through analytical evaluations. The applicant also described its flaw evaluation process, and referenced ASME Code Section XI, IWB-3640 as the code used in the evaluation process. In addition, the applicant indicated that acceptance standards provided in ASME Code Section XI, Subsections IWB, IWC, or IWD were used to determine whether a flaw is reportable.

Based on its review, the staff finds the applicant's response to RAI 4.7.5-1 acceptable because the applicant has verified that its evaluation and characterization conforms to ASME Code Section XI. Therefore, the staff's concern described in RAI 4.7.5-1 is resolved.

In LRA Section 4.7.1, the applicant referenced an NRC safety evaluation for the indication on the RCS loop C cold leg. The staff determined that additional information was required to complete its review.

In RAI 4.7.1-6, dated April 1, 2008, the staff requested that the applicant verify the dates of the SER.

In its response to RAI 4.7.1-6, dated June 2, 2008, the applicant verified that the date contained in the LRA is correct. In addition, the applicant notified the staff that it had discovered that the subject flaw evaluation was not updated using revised loading conditions derived from the extended power uprate and steam generator replacement projects. Also, the applicant noted that it had performed an assessment for the indication confirming that the evaluation would remain acceptable for 60 years. The applicant indicated that for the assessment, it used the applicable thermal transients reflecting EPU conditions; latest piping reaction loads reflecting EPU conditions, including the replacement steam replacements; and thermal aging and/or fracture toughness pursuant to NUREG/CR-4513, Revision 1.

Based on its review, the staff finds the applicant's response to RAI 4.7.1-6 acceptable because the applicant has appropriately updated its flaw evaluation to reflect the current plant conditions. Therefore, the staff's concern described in RAI 4.7.1-6 is resolved.

Based on the above review of the applicant's response to the staff's RAI question, the staff's detailed evaluation is discussed below:

In LRA Section 4.7.1, the applicant stated that it detected indications in weld DLW-LOOP3-7-S-02, the first circumferential weld after the RV nozzle-to-safe-end weld on the RCS C loop cold leg (the other end of the RV nozzle-to-safe-end is welded to an elbow). The applicant verified that the elbow was fabricated of Grade CF8M CASS and Weld DLW-LOOP3-7-S-02 was fabricated with TP 308 stainless steel. The applicant further stated that the stainless steel weld was made using a tungsten inert gas welding process for the root pass and the remaining weld passes were made using a submerged arc welding process. The applicant noted that the TLAA concerns are thermal aging of the cast material and fatigue crack growth analyses because these two issues are time-dependent.

The applicant performed three separate UT examinations of the subject weld on March 26, March 27, and March 29, 1996, respectively. The applicant also conducted a follow-up examination on the inside surface of the pipe using a driver-pickup eddy current probe to verify that the indications were not connected to the inside surface of the pipe. Examination revealed the presence of four indications grouped in a band ranging from 4 inches to 14 inches below top dead center in the 9 o'clock to 12 o'clock quadrant looking toward the vessel.

The applicant did not report Individual depths; however, the applicant stated in its April 23, 1996, flaw evaluation report to the staff that the four indications were considered bounded by a single composite flaw. The composite flaw's depth and length were 0.68 inches and 10 inches, respectively and the pipe wall thickness was 2.66 inches. Even though examinations confirmed that the indications were not connected to the pipe surface, the fracture mechanics analysis conservatively assumed the indication to be surface-breaking.

Because the subject flaw exceeded the ASME Code, Section XI, Subsection IWB-3500 acceptance criteria, the applicant performed a flaw evaluation to ensure that this indication would remain within ASME Code, Section XI, Appendix C evaluation acceptance standards. The applicant's analysis concluded that the flaw met the applicable requirements, with significant margins of safety to the end of the service life. In its safety evaluation dated May 1, 1996, the staff concluded that the reported flaw was acceptable for continued service until the end of the service life, provided the weld was reexamined during each of the next three ISI periods. In addition, the staff confirmed the presence of this flaw indication during an independent inspection on April 25, 1996. The details of its inspection are documented in the staff's Integrated Inspection Reports 50-334/96-04 and 50-412/96-04.

As required by the ASME Code, Section XI, the applicant performed successive examinations of the subject weld for the first two ISI periods (first and second periods of the third ISI Interval) and the results indicated that no measurable growth was observed since the initial detection. The third period examination should have been completed during RFO 17 (February 13 to April 19, 2006) or RFO 18 (September 24 to October 24, 2007) for Unit 1. However, the applicant inadvertently removed this examination requirement from the ISI third Interval 10-year ISI Plan for Unit 1. Upon discovering the deficiency, the applicant addressed the missing examination under the FENOC Corrective Action Program and documented its remedies in CR 08-38344. In this report, the applicant rectified the missing examination by requiring that the missed UT examination be performed in a forced outage by March 31, 2009, to comply with the 1-year period extension requirement of ASME Code, Section XI. The applicant stated that it may extend the UT examination to RFO 1R19 (April 2009) and may submit a relief request, should the examination extend beyond March 31, 2009 (*i.e.*, in the April 2009 RFO). The staff finds that the applicant has scheduled the missing UT examination and, therefore, this issue is resolved.

In a letter dated April 9, 2004, the staff approved a relief request from the applicant to use an alternative risk-Informed ISI program for the ASME Code Class 1 and 2 piping welds at Unit 1 (for the third 10-year ISI interval) and Unit 2 (for the second 10-year ISI interval). Under the staff-approved alternative risk-informed ISI program, the subject weld is not scheduled for future examinations. As for the degradation mechanism that may have caused the indication, the applicant stated that the piping is not susceptible to stress-corrosion cracking (SCC) based on the conclusions of its flaw analysis for weld DLW-LOOP3-7-S-02. The potential for SCC is minimized by assuring that materials selections are compatible with the plant operating parameters and that a corrosive environment is not present. The applicant stated that since high residual stresses, materials susceptibility, and a corrosive environment all have to be present in order to experience SCC, the materials specifications coupled with the absence of a corrosive environment assures that SCC is not a degradation mechanism for this piping. The staff noted that the subject weld is made of stainless steel and the pipe is made of CASS. Based on the operating experience of PWRs, the likelihood of both of these metals being susceptible to SCC is small.

The staff noted that in the applicant's flaw evaluation, the applicant used the fully-aged fracture toughness properties of the CASS straight pipe in lieu of the weld. The flaw is located in weld DLW-LOOP3-7-S-02, not in the pipe; therefore, the weld material properties should be used in the flaw evaluation. The applicant clarified that the fracture toughness value of the piping is lower than that of the elbow and weld and, therefore, is limiting. The staff finds that even though the indications are located in the weld, using material properties of the piping to analyze the crack growth of the indication is more conservative than using the material properties of the subject weld. The staff noted that the fracture toughness of the pipe is fully aged and is therefore acceptable for use in the flaw evaluation. Therefore, TLAA is not a concern for the thermal aging of the material because the fully-aged material properties were used in the flaw evaluation.

The transients considered in the flaw evaluation are the design transients contained in the equipment specification. The transients used in the flaw analysis were considered to be conservative, relative to the original design transients. The staff confirmed that as demonstrated in LRA Table 4.3-2, the Unit 1 original design transients bound the 60-year projected cycles. To verify the flaw growth evaluation remains valid for 60 years, the applicant will use the Metal Fatigue of Reactor Coolant Pressure Boundary Program to validate the cycles assumed in the flaw evaluation.

The staff finds that this program is acceptable to address the TLAA concern regarding the fatigue crack growth analysis.

During its response to a staff RAI, the applicant discovered that the subject flaw evaluation was not updated using loading conditions derived from the extended power uprate and steam generator replacement projects. The applicant addressed this deficiency under its Corrective Action Program. Since then, the applicant has completed a review of the original flaw evaluation previously performed for the indication at weld DLW-LOOP3-7-S-02 in accordance with the flaw evaluation procedure and acceptance criteria pursuant to ASME Code Section XI, 1989 Edition.

The applicant assessed the impact of the following items on the previous flaw evaluation results: (1) applicable thermal transients reflecting the extended power uprate conditions; (2) latest piping reaction loads reflecting extended power uprate conditions, including the replacement steam generators; and (3) thermal aging and/or facture toughness in accordance with guidance found in NUREG/CR-4513, Revision 1. The results of this assessment showed that the indication at weld DLW-LOOP3-7-S-02 would remain acceptable for the duration of plant life, including the license renewal period. The staff finds that the applicant has demonstrated that the original flaw evaluation of weld DLW-LOOP3-7-S-02 bounds the power uprate conditions.

The staff finds that the applicant has satisfactorily addressed the relevant TLAA issues for the indication detected in the RCS loop C cold leg and concludes that the indication should not be of concern for the period of extended operation. The staff also finds that the effects of aging on the intended function of the RCS Loop C cold leg will be adequately managed for the period of extended operation.

4.7.1.3 UFSAR Supplement

The applicant provided a UFSAR supplement summary description of its TLAA evaluation of the indication on the Unit 1 RCS loop C cold leg in LRA Section A.2.6.1. Based on its review of the UFSAR supplement, the staff concludes that the summary description of the applicant's actions to address the TLAA evaluation of the subject indication is adequate.

4.7.1.4 Conclusion

Based on its review, the staff concludes that pursuant to 10 CFR 54.21(c)(1)(i), the applicant has demonstrated that the flaw evaluation of the indication on the RCS loop C cold leg at Unit 1 remains valid for the period of extended operation because the applicant used the saturated CASS material properties and projected transient cycles to 60 years. The staff further concludes that, pursuant to 10 CFR 54.21(c)(1)(iii), the applicant has demonstrated that the effects of aging on the intended functions of the RCS cold leg will be adequately managed for the period of extended operation. The UFSAR supplement contains an appropriate summary description of the TLAA evaluation, as required by 10 CFR 54.21(d).

4.7.2 Reactor Vessel Underclad Cracking (Unit 1 Only)

4.7.2.1 Summary of Technical Information in the Application

In LRA Section 4.7.2, the applicant summarized its evaluation of RV underclad cracking (Unit 1 only) for the period of extended operation. Examination of Nucleoelectrica Argentina SA's Atucha-1 RV in 1970 first detected intergranular separations (underclad cracking) in low-alloy steel heat-affected zones under austenitic stainless steel weld claddings in SA-508, Class-2 RV forgings. There have been reports of these separations in SA-508, Class 2, RV forgings manufactured to a coarse-grain practice and clad by high-heat-input submerged arc processes. RG 1.43, "Control of Stainless Steel Weld Cladding of Low-Alloy Steel Components," states that detection of underclad cracks "normally requires destructively removing the cladding to the weld fusion line and examining the exposed base metal either by metallographic techniques or with liquid penetrate or magnetic particle testing methods." The WOG issued topical report WCAP-15338-A, "A Review of Cracking Associated with Weld Deposited Cladding in Operating PWR Plants," on flaw evaluations based on ASME Code Section XI to demonstrate that the Westinghouse reactor pressure vessels with underclad cracks are acceptable for 60 years.

For the Unit 2 RV, the cladding of the RV SA-508 Class 2 forgings used no high-heat-input welding processes which could induce underclad cracking; therefore, the Unit 2 RV is not susceptible to underclad cracking.

The Unit 1 RV has no SA-508, Class 2 forgings in the beltline regions. Only the vessel and closure head flanges and inlet and outlet nozzles are fabricated from SA-508, Class 2 forgings.

The WCAP-15338-A evaluation demonstrates that fatigue growth of the subject flaws will be minimal over 60 years and that the presence of underclad cracks is of no concern to the structural integrity of the RV.

The cycle assumptions in the flaw growth analysis are conservative compared to the original design cycles. LRA Table 4.3-2 shows the original design-basis transients including RCS design cycles along with the projected operational cycles that the applicant anticipates will occur for 60 years of plant life. The applicant has compared the design cycles to the 60-year projected operational cycles and determined that the design cycles are bounding for the period of extended operation. Since the applicant determined that the flaw growth analysis remains valid for 60 years using the 60-year projected operational cycles, the Metal Fatigue of Reactor Coolant Pressure Boundary Program must continue to validate the assumptions in the evaluation; therefore, disposition of the Unit 1 flaw growth TLAA complies with 10 CFR 54.21(c)(1)(i) and 10 CFR 54.21(c)(1)(iii).

4.7.2.2 Staff Evaluation

The staff reviewed LRA Section 4.7.2, to verify pursuant to 10 CFR 54.21(c)(1)(i), that the analyses remain valid for the period of extended operation and, pursuant to 10 CFR 54.21(c)(1)(iii), that the effects of aging on the intended function(s) will be adequately managed for the period of extended operation.

Intergranular cracking in low-alloy steel RV plates and forgings underneath stainless steel weld cladding (i.e., underclad cracking) has been observed for specific materials and cladding process conditions. According to RG 1.43, "Control of Stainless Steel Weld Cladding of Low-Alloy Steel Components," May 1973, underclad cracking has been reported only in RV forgings and plates of SA-508 Class 2 composition manufactured to a coarse-grain practice when clad using "high-heat-input" submerged arc welding processes. Cracking has not been observed in SA-508 Class 2 materials clad using "low-heat-input" processes, which are controlled to minimize heating of the base metal. Westinghouse Topical Report WCAP-15338-A, "A Review of Cracking Associated with Weld Deposited Cladding in Operating PWR Plants," October 2002, provides flaw evaluations for postulated underclad cracks based on the flaw evaluation guidelines in ASME Code, Section XI.

Given bounding assumptions with respect to cyclic loading, these flaw evaluations demonstrate that Westinghouse RVs with underclad cracks can operate safely and in compliance with regulatory requirements for 60 years.

In LRA Section 4.7.2, the applicant discussed its TLAA for the RV underclad cracking mechanism at Unit 1. The applicant stated that no high-heat-input welding processes which could induce RV underclad cracking were used in the cladding of the SA-508 Class 2 RV materials at Unit 2. Therefore, the Unit 2 RV was not deemed susceptible to underclad cracking. Only the RV closure head flange, inlet nozzles, and outlet nozzles for Unit 1 are fabricated from SA-508 Class 2 material subjected to high-heat-input welding processes; therefore, these items were deemed susceptible to underclad cracking. The applicant stated that the evaluation contained in WCAP-15338-A was used to demonstrate that the fatigue crack growth of any underclad cracks will be minimal over 60 years, and that the presence of the underclad cracks does not present a safety concern with respect to the structural integrity of the Unit 1 RV. The applicant provided original design cycles and 60-year projected operational cycles for Unit 1 and Unit 2 in LRA Table 4.3-2. The applicant indicated that the cycle assumptions used in the WCAP-15338-A flaw growth analysis are conservative (i.e., greater) compared to the original design cycles for the Unit 1 RCS. Furthermore, the applicant explained that it had reviewed the original design cycles against the 60-year projected operational cycles and determined that the original design cycles for Unit 1 remain bounding for the period of extended operation.

RG 1.43 states that underclad cracking has been reported only in forgings and plate material of SA-508 Class 2 composition, fabricated to a coarse-grain practice when clad using "high-heat-input" submerged arc welding processes. Underclad cracking has not been observed in SA-508 Class 2 materials clad by "low-heat-input" welding processes, which are controlled to minimize heating of the base metal. Furthermore, underclad cracking has not been observed in materials produced to a fine grain practice, regardless of the welding practice. Therefore, the staff agreed with the applicant's determination that the Unit 2 RV is not susceptible to underclad cracking because high-heat-input welding processes were not used in the cladding of the SA-508 Class 2 RV materials. For Unit 1, the staff confirmed that the original design cycles and 60-year projected operational cycles, as reported in LRA Table 4.3-2, are bounded by the cycle assumptions used in the flaw growth analysis in WCAP-15338-A. Therefore, the staff agreed with the applicant's determination that the presence of any underclad cracks in components fabricated from SA-508, Class 2 material would not significantly impact the structural integrity of the Unit 1 RV through the end of the period of extended operation.

The applicant stated that since the 60-year projected operational cycles were used to determine that the flaw growth analysis in WCAP-15338-A remains bounding for 60 years, the Metal Fatigue of Reactor Coolant Pressure Boundary program described in LRA Section B.2.27 must continue to be used to validate the assumptions for the design cycles and 60-year operational cycles used in this TLAA. The Metal Fatigue of Reactor Coolant Pressure Boundary Program is a TLAA management program that uses preventive measures to mitigate fatigue cracking caused by anticipated cyclic strains in metal components of the RCPB.
The preventive measures consist of monitoring and tracking critical thermal and pressure transients for RCS components to prevent the fatigue design limit from being exceeded. Therefore, the staff finds that the applicant's implementation of this AMP will ensure that the Unit 1 RV underclad cracking TLAA will be adequately managed for the period of extended operation, in accordance with 10 CFR 54.21(c)(1)(iii).

4.7.2.3 UFSAR Supplement

In LRA Section A.2.6.2, the applicant provided the UFSAR supplement summary description of the RV underclad cracking TLAA for Unit 1. The staff reviewed the applicant's UFSAR supplement summary description and determined that it is consistent with the RV underclad cracking TLAA in LRA Section 4.7.2. The UFSAR supplement states that the RV underclad cracking TLAA for Unit 1 will be managed through the implementation of the RCPB Metal Fatigue program through the end of the period of extended operation. The staff therefore determines that the Unit 1 UFSAR supplement summary description of the RV underclad cracking TLAA is acceptable.

4.7.2.4 Conclusion

The staff has reviewed the applicant's TLAA for RV underclad cracking, as summarized in LRA Section 4.7.2 and determines that the RV underclad cracking TLAA at Unit 1 will be managed through the applicant's implementation of the RCPB Metal Fatigue program to ensure compliance with 10 CFR 50.55a and 10 CFR Part 50, Appendix G, through the end of the period of extended operation. The staff therefore concludes that the applicant's TLAA for RV underclad cracking at Unit 1 complies with the staff's acceptance criterion for TLAAs in 10 CFR 54.21(c)(1)(iii) and that the safety margins established and maintained during the current operating term will be maintained during the period of extended operation, as required by 10 CFR 54.21(c)(1). The staff also concludes that the UFSAR supplement for Unit 1 contains an appropriate summary description of the RV underclad cracking TLAA for the period of extended operation, as required by 10 CFR 54.21(d).

4.7.3 Leak-Before-Break

4.7.3.1 Main Coolant Loop Piping Leak-Before-Break

4.7.3.1.1 Summary of Technical Information in the Application

The current LBB evaluation for the Unit 1 main coolant loop piping is documented in WCAP-11317, "Technical Justification for Eliminating Large Primary Loop Pipe Rupture as the Structural Design Basis for Beaver Valley Unit 1." By letter dated December 9, 1987, the staff approved LBB for the main coolant loop piping based on its review of WCAP-11317 (including Supplements 1 and 2). The applicant evaluated WCAP-11317 to determine whether elimination of the pipe breaks remains justified at power uprate operating conditions and whether the fracture toughness values calculated in WCAP-11317 were conservative. The LBB analyses for Unit 1 main coolant loop piping in WCAP-11317 includes CASS thermal aging and fatigue crack growth analysis. The current LBB evaluation for the Unit 2 main coolant loop piping is documented in WCAP-11923, "Technical Justification for Eliminating Large Primary Loop Pipe Rupture as the Structural Design Basis for Beaver Valley Unit 2 After Reduction of Snubbers." By letter dated April 8, 1991, the staff approved LBB for the Unit 2 main coolant loop piping based on its review of WCAP-11923. The applicant evaluated WCAP-11923 to determine whether elimination of the pipe breaks remains justified at power uprate operating conditions and whether the fracture toughness values calculated in WCAP-11923 were conservative. The LBB analyses for Unit 2 main coolant loop piping in WCAP-11923 includes CASS thermal aging and fatigue crack growth analysis.

The Unit 1 and 2 primary loop piping material is made of CASS. With CASS, thermal aging continues until the saturation or fully-aged point is reached. The LBB evaluations for both units use saturated (fully-aged) fracture toughness properties that are not material property time-dependent; thus, no further evaluation is required for license renewal. There is no thermal aging TLAA in the Unit 1 or Unit 2 main coolant loop piping LBB evaluations. Over time, accumulation of actual fatigue transient cycles could invalidate fatigue crack growth analyses. The applicant performed a fatigue crack growth analysis of the RV inlet nozzle to safe-end region to determine its sensitivity to the presence of small cracks.

The applicant selected the nozzle to safe-end connection because crack growth at this location is representative of the entire primary loop. The nozzle to safe-end connection configuration includes an SA-508 Class 2 or Class 3 stainless steel-clad nozzle connected to a stainless steel safe-end by a stainless steel (Unit 1) or nickel-based alloy (Unit 2) weld. The applicant used fatigue crack growth rate laws pursuant to ASME Code Section XI to evaluate the crack growth.

The applicant stated that the cycles used in the fatigue crack growth analyses are conservative compared to the original design cycles. LRA Table 4.3-2 shows the original design-basis transients including RCS design cycles, along with the projected operational cycles for 60 years of plant life. The applicant has determined that the design cycles are bounding for the period of extended operation. Because the 60-year projected operational cycles were used in the evaluation, the applicant will continue to use the Metal Fatigue of Reactor Coolant Pressure Boundary Program to validate the cycles assumed in the evaluation.

4.7.3.1.2 Staff Evaluation

The staff reviewed LRA Section 4.7.3.1 to verify, pursuant to 10 CFR 54.21(c)(1)(i), that the LBB analyses for the main coolant loop piping remain valid for the period of extended operation and, pursuant to 10 CFR 54.21(c)(1)(iii), that the effects of aging on the intended function of the main coolant piping will be adequately managed for the period of extended operation.

The TLAA concerns are thermal aging of the cast material and fatigue crack growth analyses of the subject piping because these two issues are time-dependent. The applicant performed the fatigue crack growth analyses based on a finite element stress analysis for the inlet nozzle safe-end region of a generic plant, typical in geometry and operational characteristics to any Westinghouse PWR System. The specific system was a plant with a piping outside diameter of 33 inches, and a wall thickness of 2.85 inches. The corresponding dimensions for Unit 1 are 34.0 inches in diameter and 3.27 inches wall thickness. The corresponding dimensions for Unit 2 are 32.46 inches in diameter and 2.5 inches wall thickness. The difference in dimensions between the typical Westinghouse plant and the Beaver Valley plants is insignificant as far as fatigue crack growth analysis is concerned. The calculated fatigue crack growth for semi-elliptic surface flaws of circumferential orientation and various depths shows that crack growth is small. The transient cycles used in the fatigue crack growth analyses are conservative compared to the original design cycles. LRA Table 4.3-2 shows the original design-basis transients, including RCS design cycles, along with the projected operational cycles for 60 years of plant life. The staff determined that the design cycles used in the fatigue crack growth analysis are bounding for 60 years of operation and that the applicant will continue to use the Metal Fatigue of Reactor Coolant Pressure Boundary Program to validate the cycles assumed in its evaluation. The staff finds that the applicant has satisfactorily addressed the TLAA concern regarding the fatigue crack growth analyses of the main coolant loop piping.

The applicant stated that the Unit 1 and Unit 2 main coolant loop piping are fabricated from CASS. Therefore, the thermal embrittlement of the CASS material could be a TLAA concern because thermal embrittlement is time-dependent. The applicant used the saturated fracture toughness for the CASS components in its LBB evaluation. The saturated fracture toughness is the lowest (the worst) fracture toughness that the CASS material can reach. Fracture toughness cannot be reduced any lower than the saturated value regardless of the time. The applicant has used saturated fracture toughness to show that main coolant piping satisfies the acceptance criteria of LBB. Therefore, the staff finds that the thermal embrittlement of the CASS piping is not a TLAA concern for Units 1 and 2.

By letter dated May 19, 2000, Christopher I. Grimes of the NRC forwarded to Douglas J. Walters of Nuclear Energy Institute, guidelines on how CASS components should be managed to minimize thermal aging of CASS components. In light of this letter, the staff asked the applicant in RAI 4.7.3-6 dated April 1, 2008, to discuss whether the CASS components in the LBB piping at Units 1 and 2 satisfy the staff guidance in its May 19, 2000 letter. The applicant clarified in a letter dated June 2, 2008 that GALL AMP XI.M12, "Thermal Aging Embrittlement of Cast Austenitic Stainless Steel (CASS)", incorporates the staff positions in its evaluation dated May 19, 2000. The applicant noted that its Thermal Aging Embrittlement of Cast Austenitic Stainless Steel (CASS) Program is a new AMP that will be consistent with GALL AMP XI.M12.

The applicant will use its Thermal Aging Embrittlement of Cast Austenitic Stainless Steel (CASS) Program to monitor the effects of loss of fracture toughness on the intended function of the component by identifying CASS materials that are susceptible to thermal aging embrittlement. For potentially susceptible materials that are part of the RCPB, the program will consist of either volumetric examination of the base metal or a component-specific flaw tolerance evaluation. The staff's evaluation of the applicants Thermal Aging Embrittlement of Cast Austenitic Stainless Steel (CASS) Program is shown in SER Section 3.0.3.1.22.

The staff finds that the applicant will use its Thermal Aging Embrittlement of Cast Austenitic Stainless Steel (CASS) Program to monitor the potential thermal embrittlement of the CASS components. The applicant also will use the saturated fracture toughness for the CASS components. Therefore, the staff finds that thermal embrittlement of the CASS material will be managed satisfactorily for the main coolant piping. Therefore, the issue raised by RAI 4.7.3-6 is resolved.

As discussed in SER Section 4.7.1, in 1996, the applicant detected an indication on the RCS loop C cold leg between the elbow and a Section of straight pipe. In RAI 4.7.3-2, dated April 1, 2008, the staff requested that the applicant discuss whether this indication is located on a segment of the pipe that has been staff-approved for LBB. If so, the staff requested that the applicant discuss whether the assumptions in the LBB analyses are still valid in light of the indication, and discuss whether the indication was fabrication or service induced.

In its response to RAI 4.7.3-2, dated June 2, 2008, the applicant confirmed that the subject flaw indication is located on a segment of the pipe (in the weld) that has been approved for LBB as shown in WCAP11317. In general, the staff-approved LBB approach excludes piping with active degradation mechanism for LBB. As stated in the *Leak-Before-Break Evaluation Procedures* published in the *Federal Register* (52 FR 32626): "Piping susceptible to IGSCC [inter-granular stress corrosion cracking] with any planar flaws in excess of the standards in IWB 3514.3 of Section XI of the ASME Code, would not be permitted to use leak-before-break analysis." The applicant stated that the piping in this case, as discussed in SER Section 4.7.1, is not susceptible to SCC based on the conclusions of the original flaw analysis for weld DLW-LOOP3-7-S-02. The potential for SCC is minimized by assuring that materials selections are compatible with the plant operating parameters and a corrosive environment is not present.

As documented in a follow-up letter dated May 1, 1996, the applicant reviewed the results of the video and eddy current examinations of the inside diameter (ID) surface of the RCS loop C cold leg. The applicant verified that there was no surface breaking indications or geometric irregularities on the ID surface. The lack of any ID surface breaking indication provides assurance that there is no need to address in-service failure mechanisms. The weld was reexamined during the first two 40-month periods (first and second periods of the third ISI Interval) following discovery, and the results indicated that no measurable growth was observed in the flaw since the initial examination. This provides additional confirmation that the indication was not a result of SCC and confirms that the LBB analysis remains valid.

The staff finds that the indication in the RCS loop C cold leg is not caused by an active degradation mechanism. Therefore, the current LBB evaluation is valid because the cold leg does not have an active degradation mechanism. On this basis, this issue raised by RAI 4.7.3-6 is resolved.

Operating experience has shown that nickel-based Alloy 600/82/182 material in the PWR environment is susceptible to primary water stress-corrosion cracking (PWSCC). In light of this emerging issue, the applicant clarified that there is no Alloy 82/182 weld metal or Alloy 600 components in the Unit 1 primary (main) loop piping. The applicant stated that the Unit 2 main coolant loop piping (RV inlet/outlet nozzles safe-end welds) and Unit 2 pressurizer surge line piping (pressurizer surge nozzle safe-end weld) contain Alloy 600/82/182 material. At this time, the applicant has no plans to perform full structural weld overlays or mechanical stress improvement of the Unit 2 main coolant inlet/outlet nozzles safe-end welds to reduce their susceptibility to PWSCC. The applicant examined the Unit 2 main coolant inlet/outlet nozzles safe-end welds using Performance Demonstration Initiative-qualified ultrasonic examination techniques in the spring of 2008. The applicant obtained greater than 90 percent ultrasonic examination coverage at all six dissimilar metal-weld locations, with no recordable indications identified.

The future inspection frequency of the Alloy 600/82/182 components will be determined by the applicant's BVPS Nickel-Alloy Nozzles and Penetrations Program. Implementation of this program is commitment 15 in the LRA Table A.4-1 and commitment 17 in Table A.5-1, for Units 1 and 2 respectively. The staff finds that the commitments in LRA Tables A.4-1 and A.5-1 are acceptable. However, the staff notes that the future inspection frequency and methods of the Alloy 600/82/182 components may be dictated by future ASME Code requirements, industry guidance, or NRC regulations for the inspection of the Alloy 600/82/182 components.

The applicant updated the current LBB analyses (WCAP-11317 and WCAP-11923) to address extended power uprate conditions on the main loop piping. The loadings, operating pressure and temperature parameters for the extended power uprate were used in the evaluation. The parameters important in the evaluation are the piping forces, moments, normal operating temperature, and normal operating pressure.

These parameters were used in the evaluation. The evaluation results show that the LBB conclusions provided in the current LBB analyses for Unit 1 and Unit 2 remain unchanged for the extended power uprate conditions. The staff confirms that the changes in the applied piping loads due to extended power uprate conditions are not sufficiently significant to change the results of the current LBB analyses.

The staff finds that the applicant has satisfactorily performed TLAA of the LBB analyses for the main coolant loop piping.

The staff also finds that the applicant has addressed the effects of PWSCC of Alloy 82/182 dissimilar metal welds and has satisfactorily evaluated the impact of the extended power uprate conditions on the subject piping; therefore, the effects of aging on the intended function of the main coolant piping will be adequately managed for the period of extended operation.

4.7.3.1.3 UFSAR Supplement

The applicant provided a UFSAR supplement summary description of its TLAA evaluation of main coolant loop piping LBB in LRA Sections A.2.6.3.1 and A.3.6.1.1. Based on its review of the UFSAR supplement, the staff concludes that the summary description of the applicant's actions to address main coolant loop piping LBB is adequate.

4.7.3.1.4 Conclusion

Based on its review, the staff concludes that pursuant to 10 CFR 54.21(c)(1)(i), the applicant has demonstrated that main coolant loop piping LBB analyses remain valid for the period of extended operation. The staff further concludes that pursuant to 10 CFR 54.21(c)(1)(iii), the applicant has demonstrated that the effects of aging on the intended function of the main coolant loop piping will be adequately managed for the period of extended operation. The UFSAR supplement contains an appropriate summary description of the TLAA evaluation of the main coolant loop piping, as required by 10 CFR 54.21(d).

4.7.3.2 Pressurizer Surge Line Piping Leak-Before-Break

4.7.3.2.1 Summary of Technical Information in the Application

The current LBB evaluation for the Unit 1 pressurizer surge line piping is documented in WCAP-12727, "Evaluation of Thermal Stratification for the Beaver Valley Unit 1 Pressurizer Surge Line." The staff approved LBB for the Unit 1 pressurizer surge line based on its May 2, 1991 review of WCAP-12727. The applicant evaluated WCAP-12727 to determine whether elimination of the Unit 1 pressurizer surge line pipe breaks from the structural design basis remains justified at power uprate operating conditions. The Unit 1 surge line piping, fabricated from wrought austenitic stainless steel, is not susceptible to reduction of fracture toughness by thermal embrittlement. Therefore, the only TLAA for Unit 1 pressurizer surge line in WCAP-12727 requiring disposition for license renewal is the fatigue crack growth analysis.

The current LBB evaluation for the Unit 2 pressurizer surge line piping is documented in WCAP-12093, "Evaluation of Thermal Stratification for the Beaver Valley Unit 2 Pressurizer Surge Line." The staff approved LBB for the Unit 2 pressurizer surge line based on its January 18, 1990 review of WCAP-12093, including Supplements 1 and 2. The LBB analyses were based on a maximum temperature difference of 315°F between the pressurizer and the hot leg. After the staff approved LBB for the Unit 2 pressurizer surge line in 1990, the plant experienced a system temperature difference of approximately 360°F during heatup. To address this issue, the applicant prepared and submitted WCAP-12093-P, Supplement 3, "Evaluation of Pressurizer Surge Line Transients Exceeding 320°F for Beaver Valley Unit 2," (proprietary). The staff approved WCAP-12093-P, Supplement 3, on April 8, 1991. The applicant concluded that this larger temperature difference does not significantly affect the maximum stress intensity, fatigue usage factor, or growth of postulated cracks nor does it impact the 40-year design life. In addition, the applicant revised its operating procedures to ensure that a 320°F system temperature difference would not be exceeded.

The applicant studied the Unit 2 pressurizer surge line LBB evaluation to determine whether elimination of pressurizer surge line pipe breaks from the structural design basis remains justified at power uprate operating conditions. The Unit 2 surge line piping, fabricated from wrought austenitic stainless steel, is not susceptible to reduction of fracture toughness by thermal embrittlement. Therefore, the only TLAA for Unit 2 pressurizer surge line in WCAP-12093 and its supplements requiring disposition for license renewal is the fatigue crack growth analysis.

The applicant's fatigue crack growth analyses of selected pressurizer surge line locations determined sensitivity to the presence of small cracks. The accumulation of actual fatigue transient cycles is the one factor in the Units 1 and 2 analyses that could be influenced by time; therefore, the fatigue crack growth analyses reported in WCAP-12727 and WCAP-12093 (including supplements) could be invalidated.

The cycle assumptions in the fatigue crack growth analyses are conservative compared to the applicant's original design cycles. LRA Table 4.3-2 shows the applicant's original design basis transients, including RCS design cycles along with the projected operational cycles for 60 years of plant life. The applicant has compared the design cycles against the 60-year projected operational cycles and has determined that the design cycles are bounding for the period of extended operation. To ensure that the 60-year projected operational cycles remain valid, the applicant will use its Metal Fatigue of Reactor Coolant Pressure Boundary Program to validate the cycles assumed in the LBB evaluation.

4.7.3.2.2 Staff Evaluation

The staff reviewed LRA Section 4.7.3.2 to verify, pursuant to 10 CFR 54.21(c)(1)(i), that the LBB analyses remain valid for the period of extended operation and pursuant to 10 CFR 54.21(c)(1)(iii), that the effects of aging on the intended function(s) will be adequately managed for the period of extended operation.

The TLAA concerns of the LBB analyses are thermal aging of the cast material and fatigue crack growth analyses. The pressurizer surge piping in Units 1 and 2 are made of wrought austenitic stainless steel which is not susceptible to thermal embrittlement. Therefore, fatigue crack growth analyses are the only TLAA of concern for Units 1 and 2 because they are time-dependent.

The accumulation of actual fatigue transient cycles is the only factor in the Units 1 and 2 LBB analyses that could be influenced by time; therefore, the fatigue crack growth analyses reported in WCAP-12727 and WCAP-12093 (including supplements), could be invalidated. The staff confirmed that a range of locations were evaluated, representing various cross sections of the surge line where thermal stratification could occur for fatigue crack growth. The circumferential positions are controlling positions because the global structural bending stress is maximum at two of the positions, while the local axial stress on the inside surface is maximum at two of the other positions. The largest initial flaw assumed to exist had a depth equal to ten percent of the wall thickness, which is the maximum acceptable flaw size pursuant to ASME Code Section XI. The wall thickness is 1.41 inches. The initial flaw depth used in the fatigue crack growth analysis is 0.141 inches. The maximum depth of crack growth after full service life was 0.347 inches, which was less than 25% of the nominal wall thickness.

The cycle assumptions in the fatigue crack growth analyses are conservative compared to the applicant's original design cycles. LRA Table 4.3-2 shows the applicant's original design basis transients, including RCS design cycles and the projected operational cycles for 60 years of plant life. The staff determines that the design cycles used in the fatigue crack growth analysis are bounding for 60 years of plant life. To ensure that the 60-year projected operational cycles remain valid, the applicant will continue to use its Metal Fatigue of Reactor Coolant Pressure Boundary Program to count the exact operational cycles to validate the assumed cycles in the LBB evaluation.

As stated above, the applicant experienced a system temperature difference of about 360°F during Unit 2 plant heatup. In RAI 4.7.3-3, dated April 1, 2008, the staff requested that the applicant clarify whether the original LBB analysis has been evaluated with a 360°F temperature difference, and whether the cycle counts in LRA Table 4.32 include this temperature transient in the LBB analysis.

In its response to RAI 4.7.3-3, dated June 2, 2008, the applicant responded that as presented in WCAP-12093-P, Supplement 3 and submitted to the staff on August 10, 1990, the characteristics of the thermal stratification transients were discussed. WCAP-12093, Supplement 3 evaluated the original LBB analysis with the addition of the observed 360°F temperature difference that occurred during a Unit 2 plant heatup. Specifically, the observed 360°F transient scenario was postulated to consist of one transient of 360°F, one transient of 342°F and three transients of 320°F. This is in addition to every design cycle developed in the original LBB analysis. The Unit 2 plant heatup is included in the operational cycle count (operational cycles as of October 15, 2003) in LRA Table 4.3-2. WCAP-12093, Supplement 3 concludes that the maximum stress intensity, fatigue usage factor, and growth of postulated cracks are not significantly affected by the observed transient of 360°F and that the design life is not affected by this larger temperature difference.

The staff confirmed that the analysis of the unanalyzed condition showed that the design life of the pipe is not affected by the 360°F temperature transient.

The applicant clarified in its response to RAI 4.7.3-4 dated June 2, 2008 that the Units 1 and 2 operating procedures were revised to include a requirement to verify that the pressurizer to C loop hot leg differential temperature is less than 320°F when the conditions to start a RCP are established during heatup. Otherwise, the RCS temperature shall be raised to reduce the differential temperature to less than 320°F prior to start of an RCP. Therefore, this operating restriction is sufficient to preclude exceeding the 320°F differential temperature design limit between the pressurizer and the hot leg during heatup.

With regard to the impact of power uprate to the pressurizer surge lines, the applicant updated the current LBB analyses (WCAP-12727 and WCAP-12093) to address extended power uprate conditions. The applicant determined the impact of the loadings and other parameters on the LBB analysis due to the extended power uprate conditions. The results of the evaluation show that all the LBB acceptance criteria and recommended margins are satisfied at the extended power uprate conditions. The staff finds that the applicant has demonstrated that the LBB evaluation of the pressurizer surge lines in Units 1 and 2 remain valid for the extended power uprate operating conditions.

The staff is also concerned with the emerging issue of PWSCC occurrence in Alloy 82/182 dissimilar metal butt welds. In its June 2, 2008 letter, the applicant confirmed that there is no Alloy 82/182 weld metal or Alloy 600 components in the Unit 1 pressurizer surge line piping. Therefore, the likelihood of PWSCC affecting the butt welds in the Unit 1 pressurizer surge line is small.

The Unit 2 pressurizer surge line does contain an Alloy 82/182 dissimilar metal weld. The applicant has installed a full structural weld overlay on the Alloy 82/182 dissimilar metal weld of the pressurizer surge line during RFO 12 (October 2 to November 12, 2006) for Unit 2. The applicant ultrasonically examined the subject Alloy 82/182 weld following structural weld overlay in the Fall of 2006. The applicant performed the examination on the required inspection volume in the weld overlay and outer 25 percent of the original dissimilar metal weld using Performance Demonstration Initiative-qualified ultrasonic examination techniques and found no recordable indications. The staff finds that the applicant has addressed PWSCC of the Alloy 82/182 dissimilar metal weld appropriately by installing the weld overlay.

The staff finds that the applicant has satisfactorily performed TLAA evaluation of the fatigue crack growth analysis and has implemented the Metal Fatigue of Reactor Coolant Pressure Boundary Program to monitor the operating cycles. The staff further finds that the effects of aging on the intended function of the pressurizer surge lines Units 1 and 2 will be adequately managed for the period of extended operation.

4.7.3.2.3 UFSAR Supplement

The applicant provided a UFSAR supplement summary description of its TLAA of the LBB analyses for the Units 1 and 2 pressurizer surge lines in LRA Sections A.2.6.3.2 and A.3.6.1.2. Based on its review of the UFSAR supplement, the staff concludes that the summary description of the applicant's actions to address TLAA of the LBB analyses for the pressurizer surge lines in Units 1 and 2 is adequate.

4.7.3.2.4 Conclusion

Based on its review, the staff concludes that pursuant to 10 CFR 54.21(c)(1)(i), the applicant has demonstrated that the LBB analysis for the Units 1 and 2 pressurizer surge lines remains valid for the period of extended operation, and pursuant to 10 CFR 54.21(c)(1)(iii), the applicant has demonstrated that the effects of aging on the intended function of the Units 1 and 2 pressurizer surge lines will be adequately managed for the period of extended operation. The UFSAR supplement contains an appropriate summary description of the TLAA evaluation of the Units 1 and 2 pressurizer surge lines, as required by 10 CFR 54.21(d).

4.7.3.3 Branch Line Piping Leak-Before-Break (Unit 2 Only)

4.7.3.3.1 Summary of Technical Information in the Application

Unit 1 has not implemented LBB on any branch line piping segments. The staff approved the LBB evaluations for the Unit 2 branch line piping in NUREG-1057, Supplement No. 4, "Safety Evaluation Report Related to the Operation of Beaver Valley Power Station Unit 2," March 1987. There are no cast materials for the subject Unit 2 piping; therefore, thermal aging of the Unit 2 branch lines is not a TLAA concern. The applicant calculated fatigue crack growth at the piping limiting locations based on a postulated conservative crack. The analysis result showed that, after a 40-year plant life, the crack would not grow to a 100% through-wall size. The fatigue transients for the crack growth evaluations were in accordance with ASME Code, Section III, Class 1 stress analyses for the particular line. As these fatigue transients and the resulting crack growth evaluation are based on a 40-year plant life, the Unit 2 branch piping LBB evaluations are TLAAs requiring disposition for the period of extended operation.

The applicant's fatigue crack growth analyses of selected RCS, RHR, and safety injection system (SIS) loop bypass line locations determined sensitivity to the presence of small cracks.

The time-dependent factor in the LBB evaluation for the Unit 2 branch lines is actual fatigue transient cycles; therefore, the fatigue crack growth analyses could be invalidated.

The applicant assumed a conservative number of cycles in its fatigue crack growth analyses compared to the Unit 1 and 2 original design cycles. LRA Table 4.3-2 shows the Units 1 and 2 original design basis transients including RCS design cycles along with the projected operational cycles for 60 years of plant life. The applicant has compared the design cycles against the 60-year projected operational cycles and determined that the design cycles are bounding for the period of extended operation. To determine the validity of the fatigue crack growth analyses for 60 years, the applicant will continue to use the Metal Fatigue of Reactor Coolant Pressure Boundary Program to validate the assumptions in the LBB evaluation.

4.7.3.3.2 Staff Evaluation

The staff reviewed LRA Section 4.7.3.3 to verify, pursuant to 10 CFR 54.21(c)(1)(i), that the applicant's LBB analyses of the Unit 2 branch lines remain valid for the period of extended operation and, pursuant to 10 CFR 54.21(c)(1)(iii), that the effects of aging on the intended functions of the branch lines will be adequately managed for the period of extended operation.

The TLAA concerns are thermal aging of the cast material and fatigue crack growth analyses of the branch lines. As stated above, the applicant has not requested the application of LBB on any Unit 1 branch lines. Therefore, the staff evaluated only the Unit 2 branch lines in terms of TLAA. The staff confirmed that there are no cast materials used in the Unit 2 branch lines. Therefore, thermal aging of the branch lines in Unit 2 is not a TLAA concern. The fatigue crack growth, which is time-dependent, is the only TLAA concern for the Unit 2 branch lines.

The staff has approved the following Unit 2 branch lines for LBB: a 10-inch RHR discharge line, a 12-inch RHR suction line, 6-inch SIS lines to hot leg and cold leg, 8-inch RCS loop bypass lines, 12-inch SIS accumulator injection lines, and the 14-inch pressurizer surge line. The applicant's LBB evaluation for the Unit 2 RCL branch line piping is documented in the *WHIPJET Program Final Report,* January 30, 1987, which includes the fatigue crack growth analysis. The staff approved the applicant's LBB evaluation in NUREG-1057, Supplement Number 4, March 1987.

The staff finds that the applicant's fatigue crack growth calculations of the Unit 2 branch lines were performed at the limiting locations based on normal and safe shutdown earthquake loads. The applicant assumed a conservative crack that exceeds the ASME Code Section XI acceptance criteria. The calculated fatigue crack growth results show that the final fatigue crack is less than the pipe wall thickness; therefore, the potential for leakage in the branch lines is small. In addition, the cycles assumed in the fatigue crack growth analyses are conservative compared to the Unit 2 original design cycles. The staff confirmed that the Unit 2 original design basis transients are bounding for the 60-year projected operational cycles. The applicant will continue to use the Metal Fatigue of Reactor Coolant Pressure Boundary Program to count the actual operating cycles in order to validate the cycle assumptions in the LBB evaluation.

The staff finds that the applicant has satisfactorily performed TLAA evaluation of fatigue crack growth calculation and has implemented the necessary AMP to monitor the operating cycles.

The applicant clarified that the LBB evaluations performed for the Unit 2 branch lines remain valid for the extended power uprate conditions based on a letter dated October 4, 2004, Pearce, L. W. (FENOC), "Beaver Valley Power Station Unit No. 1 and No. 2, BV-1 Docket No. 50-334, License No. DPR-66, BV-2 Docket No. 50-412, License No. NPF-73, License Amendment Request No. 302 and 173." The branch line piping forces and moments, operating pressure, temperature, and material properties were the important input parameters for the previous WHIPJET evaluation. The applicant evaluated the impact of extended power uprate conditions on these input parameters as one percent or less. Therefore, the staff finds that the Unit 2 branch lines are insignificantly affected and remain acceptable for extended power uprate conditions.

The staff also evaluated the branch lines with regard to the emerging issue of PWSCC in Alloy 82/182 butt welds. The applicant stated that there is no Alloy 82/182 weld metal or Alloy 600 components in the Unit 2 branch line piping (RHR, SIS and RCS loop bypass lines). Therefore, the likelihood of PWSCC affecting the dissimilar metal butt welds in the branch lines is small. The staff notes that the pressurizer surge line does contain an Alloy 82/182 butt weld. The disposition of the pressurizer surge line with respect to PWSCC is discussed in SER Section 4.7.3.2.

The staff finds that the TLAA of the LBB analyses of the Unit 2 branch lines remain valid for the period of extended operation.

The staff also finds that the effects of aging on the intended functions of the Unit 2 branch lines will be adequately managed for the period of extended operation.

4.7.3.3.3 UFSAR Supplement

The applicant provided a UFSAR supplement summary description of its TLAA of the LBB analyses for Unit 2 branch lines in LRA Section A.3.6.1.3. Based on its review of the UFSAR supplement, the staff concludes that the summary description of the applicant's actions to address the Unit 2 branch line LBB analyses is adequate.

4.7.3.3.4 Conclusion

Based on its review, the staff concludes that pursuant to 10 CFR 54.21(c)(1)(i), the applicant has demonstrated that the LBB analyses of the Unit 2 branch line remain valid for the period of extended operation and that pursuant to 10 CFR 54.21(c)(1)(iii), the applicant has also demonstrated that the effects of aging on the intended function of the Unit 2 branch lines will be adequately managed for the period of extended operation. The UFSAR supplement contains an appropriate summary description of the TLAA evaluation of the Unit 2 branch lines, as required by 10 CFR 54.21(d).

4.7.4 High-Energy Line Break Postulation

4.7.4.1 Summary of Technical Information in the Application

In LRA Section 4.7.4, the applicant summarized its evaluation of High Energy Line Break (HELB) postulation for the period of extended operation. In accordance with 10 CFR 50, Appendix A, GDC 4, "Environmental and Dynamic Effects Design Bases," the applicant has taken special measures in the design and construction of Units 1 and 2 to protect systems, structures or components required to place the reactor in a safe cold shutdown condition from the dynamic effects of postulated rupture of piping.

The applicant stated that as described in UFSAR Section 5.2.6.3 for Unit 1, specific placement of piping and components (protection barriers) ensures compliance with this criterion. The careful layout of piping and components offers adequate protection against the dynamic effects of a postulated pipe rupture, except in the main steam and feedwater lines outside the crane wall. For these two piping systems, the applicant used RG 1.46, "Protection Against Pipe Whip Inside Containment," as the base document for evaluation of the break locations. The applicant designed the Unit 1 piping systems in accordance with ANSI B31.1 standards, which is not addressed in RG 1.46. However, ANSI B31.1 is similar to ASME Code Section III, Class 2 piping; therefore, the applicant used the RG 1.46 Class 2 piping requirements for these lines. Similarly, the applicant states in UFSAR Section 3.6B.2.1.1.1 for Unit 2 that determination of the break locations for ASME Code Section III, Classes 1, 2, and 3 piping systems (outside the scope of those exempted through LBB evaluations as described in LRA Section 4.7.3) is in compliance with RG 1.46.

The applicant also stated that the ANSI B31.1, ASME Code Class 2, and ASME Code Class 3 postulated break locations are determined based on where the maximum stress range derived from the piping stress analysis from normal, upset, and OBE conditions exceeds established criteria. These postulated break location determinations do not use CUFs; therefore, they require no further evaluation for license renewal.

The applicant further stated that for the Unit 2 Class 1 systems, RG 1.46 determines postulated break locations, in part, by using any intermediate locations between terminal ends, where the CUF derived from the piping fatigue analysis under loadings from specified seismic events and operational plant conditions exceeds 0.1. These fatigue evaluations are TLAAs based on a set of fatigue transients for the life of the plant. The cycle assumptions in the fatigue crack growth analyses are conservative compared to the original design cycles. LRA Table 4.3-2 shows the original design-basis transients including RCS design cycles along with the projected operational cycles for 60-years of plant life. The applicant has compared the design cycles to the 60-year projected operational cycles and has determined that the design cycles are bounding for the period of extended operation. The applicant concluded that based on its determination that the fatigue crack growth analyses remain valid for 60-years using the 60-year projected operational cycles, the Metal Fatigue of Reactor Coolant Pressure Boundary Program must continue to validate the assumptions in the evaluation. Therefore, the applicant concluded that the piping fatigue analyses for determining postulated break locations in Class 1 lines remain valid for the period of extended operation, in accordance with 10 CFR 54.21(c)(1)(i) and 10 CFR 54.21(c)(1)(iii).

4.7.4.2 Staff Evaluation

The staff reviewed LRA Section 4.7.4 to verify, pursuant to 10 CFR 54.21(c)(1)(i), that the analyses remain valid for the period of extended operation and, pursuant to 10 CFR 54.21(c)(1)(iii), that the effects of aging on the intended function(s) will be adequately managed for the period of extended operation.

In LRA Section 4.7.4, the staff reviewed the applicant's technical information and onsite documentation supporting the applicant's conclusion that the analysis postulation of the HELB remains valid for the period of extended operation. The staff also interviewed the applicant's technical staff to verify the description of the LRA and its supplementing documents.

The staff noted the criteria for protection against dynamic effects associated with a major pipe rupture is described in UFSAR Section 5.2.6.3 for Unit 1. The staff further noted that the piping in Unit 1 was designed in accordance with the ANSI B31.1 code. For such piping, postulated break locations is dependent, in part, on the maximum stress range associated with normal and upset conditions and an OBE event derived from the piping stress analysis, and are not dependent on CUFs. The staff further noted that the Unit 1 HELB postulation for license renewal does not meet the definition of a TLAA as described in 10 CFR 54.3(a); therefore, a TLAA evaluation is not required.

For Unit 2, the design basis of HELB postulation is provided in UFSAR Section 3.6.

The staff noted in the description in LRA Section 4.7.4 the applicant stated, "The cycle assumptions used in the fatigue crack growth analyses are conservative compared to the BVPS original design cycles," and "Since the 60-year projected operational cycles were used in determining that fatigue crack growth analyses remains valid for 60 years, the Metal Fatigue of Reactor Pressure Boundary Program must continue to be used to validate the assumptions used in the evaluations." In RAI 4.7.4-1 dated May 28, 2008, the staff requested that the applicant explain the reason why fatigue crack growth analyses are evaluated for the HELB postulation TLAA.

In its response to RAI 4.7.4-1, dated July 11, 2008, the applicant stated that use of the term "fatigue crack growth analyses" was a typographical error. Therefore the applicant amended LRA Section 4.7.4 to use the correct terminology, "fatigue analyses." The staff concludes that the applicant's amendment uses the correct terminology of "fatigue analyses", as opposed to "fatigue crack growth analyses." Therefore, the staff's concern described in RAI 4.7.4-1 is resolved.

The staff noted in LRA Section 4.7.4 that the applicant utilized the projected operational cycles expected to occur in 60 years of operation. The applicant compared the projected cycles with the design cycles and determined that the design cycles are bounding. During its review, the staff noted that additional clarification was required for the Class 1 high-energy piping. In RAI 4.7.4-2, dated May 28, 2008, staff requested that the applicant confirm whether any Class 1 high-energy piping locations with a CUF of less than 0.1 by the CLB may exceed the 0.1 CUF during the period of extended operation.

In its response to RAI 4.7.4-2, dated July 11, 2008, the applicant stated that the design CUF for Class 1 high-energy piping locations has not increased; therefore, there are no locations where the design CUF had been less than 0.1 or will be greater than 0.1.

Based on its review, the staff finds the applicant's response acceptable because the applicant has confirmed that there are no new rupture locations that are postulated, the staff finds the applicant's response to RAI 4.4.7-2 acceptable. Therefore, the staff's concern described in RAI 4.4.7-2 is resolved.

Based on its review and the applicant's confirmation as described above, the staff determines that the design CUF is bounding for the period of extended operation and the applicant has demonstrated the analyses remain valid for the period of extended operation in accordance with 10 CFR 54.21(c)(1)(i). Since the applicant has utilized the 60-year projected operational cycles in determining the fatigue analyses remains valid for the period of extended operation, the applicant's Metal Fatigue of Reactor Coolant Pressure Boundary Program must validate the assumptions in these evaluations. Therefore the staff finds that the effects of aging on the intended functions will be adequately managed for the period of extended operation in accordance with 10 CFR 54.21(c)(1)(iii).

4.7.4.3 UFSAR Supplement

The applicant provided a UFSAR supplement summary description of its TLAA evaluation of HELB postulation in LRA Section A.3.6.2. Based on its review of the UFSAR supplement, the staff concludes that the summary description of the applicant's actions to address HELB postulation is adequate.

4.7.4.4 Conclusion

Based on its review as discussed above, the staff concludes that the applicant has provided an acceptable demonstration that the postulated break location in Class 1 lines for Unit 1 does not meet the definition of a TLAA in accordance with 10 CFR 54.3(a). The staff further concludes that the applicant has provided an acceptable demonstration for HELB TLAA for Unit 2, and pursuant to 10 CFR 54.21(c)(1)(i), that the analyses remain valid for the period of extended operation and, pursuant to 10 CFR 54.21(c)(1)(iii), that the effects of aging on the intended function(s) will be adequately managed for the period of extended operation. The staff also reviewed the UFSAR supplement for this TLAA and concludes that it provides an adequate summary description of the HELB TLAA evaluation, as required by 10 CFR 54.21(d).

4.7.5 Settlement of Structures (Unit 2 Only)

4.7.5.1 Summary of Technical Information in the Application

In LRA Section 4.7.5, the applicant stated that settlement of Unit 2 Category I structures and buried piping is documented in UFSAR Section 2.5.4.13. The applicant's general analysis and evaluation of structural settlement is documented in UFSAR Section 2.5.4.10.2. The applicant monitored the settlement of the structures and buried piping during construction and will continue to do so throughout the life of the plant, until it determines that the settlement for a particular structure is stable; after which, it will discontinue monitoring. The applicant determined the stability of settling structures for Unit 2 pursuant to the Settlement Monitoring Program (Unit 2 only) description in LRA Section B.2.37. This program provides the requirements to measure the settlement of Unit 2 structures at selected locations. If the measured settlements of a structure exceed the expected settlements, the applicant stated that it will review the current settlement analysis as it relates to the integrity of the structure and the maintenance of settlement assumptions made in the piping stress analyses.

The applicant also stated that the Settlement Monitoring Program (Unit 2 only) ensures that the current 40-year settlement assumptions in the Unit 2 buried piping stress analyses are maintained for the period of extended operation. Therefore, the Unit 2 piping fatigue TLAAs have been disposition in accordance with 10 CFR 54.21(c)(1)(iii).

4.7.5.2 Staff Evaluation

The staff reviewed LRA Section 4.7.5 to verify pursuant to 10 CFR 54.21(c)(1)(iii), that the effects of aging on the intended function(s) will be adequately managed for the period of extended operation.

The staff reviewed LRA Section 4.7.5 for Unit 2 for correlation with the provisions in UFSAR Section 2.5.4.10.2 and determined that additional information was needed to complete its review.

In RAI 4.7.5-1, dated April 1, 2008, the staff requested that the applicant indicate, for each structure included in the Settlement Monitoring Program, whether the structure is currently being monitored and if it will be monitored to the end of the period of extended operation.
In its response to RAI 4.7.5-1, dated April 30, 2008, the applicant stated that it monitors all Unit 2 structures under the Settlement Monitoring Program (Unit 2 only) to the end of the period of extended operation. Each Category I structure in the program has been monitored since plant construction and will be monitored throughout the life of the plant, until its settlement stability has been determined. Settlement monitoring will be discontinued for structures that meet the criteria for stability.

The applicant further stated that settlement is measured by markers placed in the vicinity of the structures. A marker is considered stable if a trend can be established over a reasonable time frame (two to three years) that shows the marker has achieved and maintained a "fixed" elevation within a tolerance of plus or minus 1/8 inch. The applicant noted that all Unit 2 safety-related structures were determined to be stable with the exception of the following three locations: (1) refueling water storage tank pad and shield wall; (2) safeguards area building; and (3) valve pit. The applicant stated that it will continue to monitor these structures until a fixed elevation is achieved, pursuant to the Settlement Monitoring Program (Unit 2 only).

The staff finds the applicant's response to RAI 4.7.5-1 acceptable because it conforms to current industry practice. Therefore, the staff's concern described in RAI 4.7.5-1 is resolved.

In RAI 4.7.5-2, dated April 1, 2008, the staff requested that the applicant provide a list of safety-related piping systems that are subject to differential settlement of the attached structures. For each system, the staff requested the projected 40-year and 60-year differential settlement of the anchor points, and the highest projected stresses for 60-year operation.

In its response, dated April 30, 2008, the applicant stated that no safety-related piping is monitored for differential settlement under the Settlement Monitoring Program (Unit 2 only), as described in UFSAR Section 2.5.4.13 for Unit 2. The applicant went on to state that if the settlements at selected locations of a Unit 2 structure exceed the anticipated settlements, it will review the current analysis as it relates to the integrity of the structure and the assumed settlements in the associated piping stress analyses, in accordance with the provisions of the Settlement Monitoring Program (Unit 2 only).

The staff reviewed the applicant's response to RAI 4.7.5-2 and finds it acceptable, because the integrity of safety-related structures and safety-related piping in Unit 2 is based on provisions of the Settlement Monitoring Program (Unit 2 only) which was previously accepted by the staff. Therefore, the staff's concern described in RAI 4.7.5-2 is resolved.

In RAI 4.7.5-3, dated April 1, 2008, the staff requested that the applicant list the Unit 2 structures that were initially monitored under the Settlement Monitoring Program (Unit 2 only) and are no longer monitored, and provide the basis for removing these structures from the monitoring program.

In its response to RAI 4.7.5-3, dated April 30, 2008, the applicant stated that the following safety-related structures were judged to have achieved settlement stability, by meeting the criterion that marker elevation remain within a plus or minus tolerance band of 0.125 inches for a period of two to three years: (1) auxiliary building; (2) diesel generator building; (3) emergency outfall structure; (4) fuel and decontamination building; (5) primary plant demineralized water tank pad and enclosure; (6) reactor containment building; (7) Unit 2 control room extension; (8) service building; (9) main steam and cable vault; and (10) intake structure. The applicant stated that it had discontinued monitoring of these buildings once it had determined that the settlement markers were stable.

The staff finds the applicant's response to RAI 4.7.5-3 acceptable because the applicant has demonstrated that settlement of the previously identified structures has stabilized, and has provided justification for discontinuing of monitoring of these structures for the period of extended operation. Therefore, the staff's concern described in RAI 4.7.5-3 is resolved.

The last sentence of LRA Section 4.7.5 indicates that the Unit 2 piping fatigue TLAAs have been disposition in accordance with 10 CFR 54.21 (c)(1)(iii). In RAI 4.7.5-4, dated April 1, 2008, the staff requested that the applicant identify the fatigue effects on the buried piping associated with piping settlement, and to clarify how the piping fatigue TLAA for the buried piping was dispositioned.

In its response to RAI 4.7.5-4, dated April 30, 2008, the applicant stated that the use of the term "fatigue" was incorrect, as there is no fatigue associated with buried piping. The term stress should have been used instead of the term "fatigue." The last sentence should have read: "Therefore, the TLAAs associated with the Unit 2 piping stress analysis have been dispositioned in accordance with 10 CFR 50.54 (c)(1)(iii)." The applicant has submitted Amendment 7 to revise LRA Section 4.7.5.

The staff finds the applicant's response to RAI 4.7.5.-4 acceptable because the applicant clarified the language in the last sentence in LRA Section 4.7.5 and submitted a revision (Amendment 7). Therefore, the staff's concern described in RAI 4.7.5-4 is resolved.

4.7.5.3 UFSAR Supplement

The applicant provided a UFSAR supplement summary description of its TLAA of settlement of structures for Unit 2 in LRA Section A.3.6.3.

By letter dated April 30, 2008, the applicant revised this Section of the supplement to state that "the TLAAs associated with the Unit 2 piping stress analysis have been dispositioned in

4-91

accordance with 10 CFR 50.54 (c)(1)(iii)." The staff finds this acceptable since this statement corresponds to the revision to LRA Section 4.7.5, as discussed in the applicant's response to RAI 4.7.5-4.

Based on its review of the UFSAR supplement, the staff concludes that the summary description of the applicant's actions to address the Unit 2 settlement of structures is adequate.

4.7.5.4 Conclusion

The staff has reviewed the applicant's submittal, in accordance with SRP-LR Section 4.7.3.1 and finds that the settlement AMP provides assurance that the provisions of UFSAR Section 2.5.4.10.2 will be extended to the end of the period of extended operation. The staff concludes that the applicant's commitment to continue monitoring the effects of Category I structure settlement on buried piping attached to these structures will provide assurance that the stresses in buried pipes resulting from the differential settlement of the attached Category I buildings will not be exceeded for the life of the plant.

Based on its review, the staff concludes that, pursuant to 10 CFR 54.21(c)(1)(iii), the applicant has demonstrated that for the Unit 2 Settlement TLAA, the effects of aging on the intended function(s) of the attached buried piping will be adequately managed for the period of extended operation. The staff also concludes that the UFSAR supplement contains an appropriate summary description of this TLAA evaluation for the period of extended operation, in accordance with the requirements of 10 CFR 54.21(d).

4.7.6 Crane Load Cycles

4.7.6.1 Summary of Technical Information in the Application

In LRA Section 4.7.6, the applicant summarized its evaluation of crane load cycles for the period of extended operation. Generic Letter (GL) 80-113, "Control of Heavy Loads," requests that all licensees of operating plants review their controls for handling heavy loads for the extent to which these controls satisfy the guidelines of NUREG-0612, "Control of Heavy Loads at Nuclear Power Plants," and for changes and modifications required to fully satisfy the guidelines. NUREG-0612 requires verification that crane designs comply with the guidelines of Crane Manufacturers Association of America Specification #70 (CMAA-70), "Specifications for Electric Overhead Traveling Cranes," and ANSI B30.2-1976, Chapter 2-1, "Overhead and Gantry Cranes," and a demonstration of equivalency of actual design requirements not in specific compliance with these standards. CMAA-70 Section 3.4.8 requires the crane design determination of allowable stress range for repeated loads. Allowable stress range is based on service class and joint category. The service class is based on a calculation of mean effective load factor (includes load magnitude and load probability), load classes, and load cycles. The minimum number of CMAA-70 load cycles is 20,000 for Class A cranes with a mean effective load factor range of 0.35-0.53. The service class load cycle parameter is based on the estimated number of load cycles (crane lifts) over the service life of the component and is, therefore, a TLAA in accordance with 10 CFR 54.3.

The staff's review of the Unit 1 response to GL 80-113 was published by the NRC in a technical evaluation report, "Control of Heavy Loads." As stated in the report, the following two Unit 1 cranes are designed to CMAA-70 standards and require TLAAs:

- Fuel cask crane (CR-15); and,
- Moveable platform and hoists crane (CR-27).

The applicant, in its September 21, 1981 response to the guidance in NUREG-0612 and GL 80-113, determined that the following three Unit 2 cranes are designed to CMAA-70 standards and require TLAAs:

- Polar crane (CRN201);
- Spent fuel cask trolley (CRN215); and,
- Moveable platform with hoists crane (CRN227).

The applicant stated that the two Unit 1 and the three Unit 2 cranes may be conservatively classified as Service Class A cranes. Conservatively estimated total load cycles and mean effective load factors for the five cranes for the period of extended operation are well below 20,000, with mean effective load factors maintained within Service Class A bounds (0.35 - 0.53) for 60 years. The applicant therefore concluded that the crane allowable stress ranges pursuant to CMAA-70 will remain valid through the period of extended operation in accordance with 10 CFR 54.21(c)(1)(i).

4.7.6.2 Staff Evaluation

The staff reviewed LRA Section 4.7.6 to verify, pursuant to 10 CFR 54.21(c)(1)(i), that the analyses remain valid for the period of extended operation.

In LRA Section 4.7.6, the applicant stated that total load cycles are well below 20,000 and mean effective load factors are maintained within or below the Service Class A bounds (0.35 - 0.53) for 60 years. However, the applicant did not provide any information on how the load cycles were calculated to conclude that the stress ranges remain valid through the period of extended operation. In RAI 4.7.6-1, dated May 22, 2008, the staff requested that the applicant provide the projected number of cycles calculated for 60 years for each of these cranes.

In its response to RAI 4.7.6-1, dated July 24, 2008, the applicant provided the following information in a table showing the 60-year projected crane cycles and how they were calculated. The applicant assumed 36 and 39 projected RFOs for Units 1 and 2, respectively.

The Unit 1 spent fuel cask crane, rated for 125 tons, is used to lift spent fuel shipping cask weighing 21.5 tons. At 2.1 lifts per outage, the projected number of cycles is 75. The staff concurs with the applicant's estimate of 75 cycles and finds it to be less than the CMAA 70 limit of 20,000 cycles by several orders of magnitude. Therefore, the staff finds that the Unit 1 spent fuel cask crane was adequately evaluated for the period of extended operation.

The Unit 1 moveable platform and hoists cranes, rated for 5 tons each, are used for fuel assembly movements weighing 2.5 tons each. The applicant calculated 502 lifts per outage based on full core offload and onload, with additional fuel assembly reshuffles and new fuel assemblies. The projected cycles based on 502 lifts are about 18,150. The staff concurs with this estimate and finds it to be less than the CMAA 70 limit of 20,000 cycles. Therefore, the staff finds that the Unit 1 moveable platform and hoists cranes were adequately evaluated for the period of extended operation.

The Unit 2 polar crane, rated for 167 tons is used to lift RV head and attachment weighing 134 tons, and other equipment weighing significantly less. The projected cycles was calculated at 2,088. The staff concurs with the applicant's estimate of 2,088 cycles and finds it to be less than the CMAA 70 limit of 20,000 cycles by several orders of magnitude. Therefore, the staff finds that the Unit 2 polar crane was adequately evaluated for the period of extended operation.

The Unit 2 spent fuel cask trolley, rated for 125 tons, is used to lift spent fuel shipping cart weighing 100 tons. The projected cycles was calculated at 206. The staff concurs with the applicant's estimate of 206 cycles and finds it to be less than the CMAA 70 limit of 20,000 cycles by several orders of magnitude. Therefore, the staff finds that the Unit 2 spent fuel cask trolley was adequately evaluated for the period of extended operation.

The Unit 2 moveable platform and hoists cranes, rated for 10 tons each, are used for fuel assembly movements weighing 3 tons each. The applicant calculated 502 lifts per outage based on full core offload and onload, with additional fuel assembly reshuffles and new fuel assemblies. The projected cycles based on 502 lifts are about 19,800. The staff concurs with this estimate and finds it to be less than the CMAA 70 limit of 20,000 cycles. Although the number of cycles is very close to the CMAA 70 limit, the load lifted is 30% of the rated capacity of the cranes.

Therefore, the staff finds that the Unit 2 moveable platform and hoists cranes were adequately evaluated for the period of extended operation.

The staff finds the applicant's response to RAI 4.7.6-1 acceptable because the applicant has provided the requested crane load cycle calculations that adequately demonstrate that the crane load stress ranges remain valid through the period of extended operation. Therefore, the staff's concern described in RAI 4.7.6-1 is resolved.

4.7.6.3 UFSAR Supplement

The applicant provided a UFSAR supplement summary description of its TLAA evaluation of crane load cycles in LRA Sections A.2.6.4 and A.3.6.4. Based on its review of the UFSAR supplement, the staff concludes that the summary description of the applicant's actions to address crane load cycles is adequate.

4.7.6.4 Conclusion

Based on its review, as discussed above, the staff concludes that the applicant has demonstrated, pursuant to 10 CFR 54.21(c)(1)(i), that, for crane load cycles, the analyses remain valid for the period of extended operation.

The staff also concludes that the UFSAR supplement contains an appropriate summary description of the TLAA evaluation, as required by 10 CFR 54.21(d).

4.8 Conclusion for TLAAs

The staff reviewed the information in LRA Section 4, "Time-Limited Aging Analyses." Based on its review, the staff concludes that the applicant has provided a sufficient list of TLAAs, as defined in 10 CFR 54.3 and that the applicant has demonstrated that: (1) the TLAAs will remain valid for the period of extended operation, as required by 10 CFR 54.21(c)(1)(i); (2) the TLAAs have been projected to the end of the period of extended operation, as required by 10 CFR 54.21(c)(1)(ii); or (3) that the effects of aging on intended function(s) will be adequately managed for the period of extended operation, as required by 10 CFR 54.21(c)(1)(iii). The staff also reviewed the UFSAR supplement for the TLAAs and finds that the supplement contains descriptions of the TLAAs sufficient to satisfy the requirements of 10 CFR 54.21(d). In addition, the staff concludes, as required by 10 CFR 54.21(c)(2) that no plant-specific, TLAA-based exemptions are in effect.

With regard to these matters, the staff concludes that there is reasonable assurance that the activities authorized by the renewed licenses will continue to be conducted in accordance with the CLB, and that any changes made to the CLB, in order to comply with 10 CFR 54.29(a), are in accordance with the Atomic Energy Act of 1954, as amended, and NRC regulations.

SECTION 5

REVIEW BY THE ADVISORY COMMITTEE ON REACTOR SAFEGUARDS

In accordance with Title 10, Part 54, of the *Code of Federal Regulations*, the Advisory Committee on Reactor Safeguards (ACRS) will review the license renewal application (LRA) for Beaver Valley Power Station, Units 1 and 2. The ACRS Subcommittee on Plant License Renewal will continue its detailed review of the LRA after this safety evaluation report (SER) is issued. FirstEnergy Nuclear Operating Company, Inc. (the applicant) and the staff of the United States (US) Nuclear Regulatory Commission (NRC) (the staff) will meet with the subcommittee and the full committee to discuss issues associated with the review of the LRA.

The NRC staff issued its safety evaluation report (SER) with open item related to the renewal of operating license for Beaver Valley Power Station, Units 1 and 2 on January 9, 2009. On February4, 2009, the applicant presented its license renewal application, and the staff presented its review findings to the ACRS Plant License Renewal Subcommittee. The staff reviewed the applicant's comments on the SER and completed its review of the license renewal application. The staff's evaluation is documented in an SER that was issued by letter dated June 8, 2009.

During the 565th meeting of the ACRS on September 11, 2009, the ACRS completed its review of the Beaver Valley Power Station, Units 1 and 2 license renewal application and the staff's SER. The ACRS documented its finding in a letter to the Commission dated September 16, 2009. A copy of this letter is provided on the following pages of this SER section.

**UNITED STATES
NUCLEAR REGULATORY COMMISSION
ADVISORY COMMITTEE ON REACTOR SAFEGUARDS
WASHINGTON, DC 20555 - 0001**

September 16, 2009

The Honorable Gregory B. Jaczko
Chairman
U.S. Nuclear Regulatory Commission
Washington, DC 20555-0001

SUBJECT: REPORT ON THE SAFETY ASPECTS OF THE LICENSE RENEWAL
                        APPLICATION FOR THE BEAVER VALLEY POWER STATION, UNITS 1 AND 2

Dear Chairman Jaczko:

During the 565th meeting of the Advisory Committee on Reactor Safeguards,
September 10-12, 2009, we completed our review of the license renewal application for the
Beaver Valley Power Station (BVPS), Units 1 and 2, and the final Safety Evaluation Report
(SER) prepared by the NRC staff. We also reviewed this matter during our 564th meeting on
July 8-10, 2009, and completed a report. The issuance of the report was delayed pending
review of new information submitted by the applicant, FirstEnergy Nuclear Operating Company
(FENOC), and the associated Supplemental SER prepared by the staff. Our Plant License
Renewal Subcommittee also reviewed this matter during its meeting on February 4, 2009.
During these reviews, we had the benefit of discussions with representatives of the NRC staff
and FENOC. We also had the benefit of the documents referenced. This report fulfills the
requirement of 10 CFR 54.25 that the ACRS review and report on all license renewal
applications.

CONCLUSIONS AND RECOMMENDATIONS

1. The programs established and committed to by the applicant to manage age-related
   degradation, including planned supplemental visual and volumetric examinations of the
   containment liners, provide reasonable assurance that BVPS, Units 1 and 2, can be
   operated in accordance with its current licensing basis for the period of extended operation
   without undue risk to the health and safety of the public.

2. The impact of containment liner corrosion on the current licensing basis of the plant is being
   reviewed and will be resolved under the provisions of the applicant's current 10 CFR Part 50
   operating licenses.

3. The FENOC application for renewal of the operating licenses of BVPS, Units 1 and 2,
   should be approved.

BACKGROUND AND DISCUSSION

BVPS consists of two Westinghouse 3-loop pressurized water reactors with subatmospheric containments (originally operated at 10 psia, now at about ½ psi below atmospheric) and is located on the south bank of the Ohio River in the Borough of Shippingport in Beaver County, Pennsylvania, approximately 25 miles northwest of Pittsburgh. The current licensed power rating of each of the BVPS units is 2,900 megawatts thermal with a gross electrical output of approximately 974 megawatts for Unit 1 and 969 megawatts for Unit 2. FENOC requested renewal of the BVPS, Units 1 and 2 operating licenses for 20 years beyond the current license terms, which expire on January 29, 2016 for Unit 1, and May 27, 2027 for Unit 2.

In the final SER, the staff documented its review of the license renewal application and other information submitted by the applicant or obtained from the staff audit and inspection at the plant site. The staff reviewed the completeness of the applicant's identification of the structures, systems, and components (SCCs) that are within the scope of license renewal; the integrated plant assessment process; the applicant's identification of the plausible aging mechanisms associated with passive, long-lived components; the adequacy of the applicant's Aging Management Programs (AMPs); and the identification and assessment of time-limited aging analyses (TLAAs) requiring review.

In the BVPS license renewal application, FENOC identified the SSCs that fall within the scope of license renewal. For these SSCs, the applicant performed a comprehensive aging management review. The final SER identifies 35 commitments for Unit 1 and 36 for Unit 2, as well as three license conditions for both units.

The BVPS application either demonstrates consistency with the Generic Aging Lessons Learned (GALL) Report or documents deviations to the specified approaches in this Report. The application includes very few exceptions, being consistent with 92% of aging management review line items specified in the GALL Report. We reviewed the exceptions and agree with the staff that they are acceptable.

The staff conducted a license renewal audit and inspection at BVPS. The audit verified the appropriateness of the scoping and screening methodology, AMPs, aging management review, and TLAAs. The site inspection verified that the license renewal requirements are appropriately implemented. Based on the audit and inspection, the staff concluded in the final SER that the proposed activities will adequately manage the effects of aging of SSCs identified in the application and that the intended functions of these SSCs will be maintained during the period of extended operation. We agree with this conclusion.

During its site inspection, the staff observed water in manholes that contain medium-voltage cables that are important to safety. The applicant has agreed that, although the cables may be suitable for submerged service, they are not qualified for that service. They have made commitments to demonstrate, using an acceptable methodology, that the cables will continue to perform their intended function; or will implement measures to minimize cable exposure to significant moisture; or will replace the cables with cables qualified for submerged service.

The applicant identified the systems and components requiring TLAAs and reevaluated them for the period of extended operation. The staff concluded that the applicant has provided an adequate list of TLAAs. Further, the staff concluded that the applicant has met the requirements of the License Renewal Rule by demonstrating that the TLAAs will remain valid for the period of extended operation, or that the TLAAs have been projected to the end of the period of extended operation, or that the aging effects will be adequately managed for the period of extended operation.

Staff reviews of operating experience have identified liner corrosion as an issue challenging containment integrity. Two separate instances of corrosion attack were discovered at BVPS, Unit 1, one in 2006 and one in 2009. These discoveries raised questions as to whether corrosion between the liner and the concrete is no longer active or will continue as the plant ages.

The 2006 discovery occurred when a temporary construction opening was made for the replacement of the Unit 1 steam generators and reactor vessel head. Degradation was observed on the inaccessible side of the steel liner. Analyses and evaluations of the Unit 1 containment liner corrosion were performed for FENOC by several contractors, including FirstEnergy Beta Laboratory and Shaw Stone & Webster, Inc.

Shaw Stone & Webster, Inc., evaluated the condition of the Unit 1 containment liner regarding the extent of the degradation and effects on its intended function as a leak tight membrane. The evaluation included consideration of the impact of an additional 20 years of operation as a result of license renewal on the recurring Integrated Leak Rate Test loading.

It was concluded that the degradation was pitting corrosion with no evidence of stress corrosion or microbiological attack. The corrosion occurred after welding and construction of the liner plate because the corrosion pitting was even across the weld, the heat affected zone of the base material, and both edges of the weld. If the corrosion had occurred prior to construction, there would be uneven corrosion across these areas due to the weld preparation and the welding process.

Approximately 1% of the observable liner plate contained corroded areas and a much smaller percentage of the rebar surface area had evidence of corrosion. The analysis concluded that the concrete did not contain corrosive agents and that no general corrosion is active in the area between the liner plate and the concrete.

The staff finds that the applicant has adequately explained the observed corrosion of the liner plate and that there is no active mechanism for corrosion. The staff agrees that the degraded conditions found on the liner in 2006 did not adversely affect its mechanical and/or structural function as a leak-tight membrane.

Following the 2006 discovery, the containment inspection procedures for Units 1 and 2 were modified to include: when paint or coatings are removed for further inspection, the paint or coatings shall be visually examined by a qualified VT-3 inspector prior to removal; and if the visual examination detects surface flaws on the liner or suspect areas on the liner plate that could potentially impact the leak tightness or structural integrity of the liner, then surface or volumetric examinations shall be performed to characterize the degradation. Staff agrees that these additional examination requirements and the use of the FENOC Corrective Action Program provide reasonable assurance that potential corrosion on the concrete side of the containment liner plate will be identified and managed.

On April 23, 2009, during a Unit 1 IWE inspection, i.e., visual inspection of 100% of all accessible portions of the containment steel liner, a paint blister was discovered on the containment liner. Further investigation revealed a rectangular through-wall hole in the containment liner, approximately 1" x 3/8". Subsequently, ultrasonic measurements were taken in the vicinity surrounding the defect to determine the extent of liner thinning. These measurements revealed indications of localized type corrosion. As a result, the applicant removed a 2 inch by 5 inch portion of the affected liner plate to further evaluate and characterize the condition.

Removal of the degraded liner section revealed a partially decomposed piece of wood embedded in the concrete containment wall, located at the interface with the steel liner plate directly behind the through-wall liner hole. Laboratory analysis indicated that the wood contained approximately 13% moisture and low pH of 3.7, i.e., mildly acidic. The applicant determined that such conditions were sufficient to promote the corrosion mechanism and cause the through-wall flaw in the liner over time, i.e., since construction in the early 1970s.

As a result of the 2009 event, visual examinations of 100% of the accessible liner area have been scheduled for the Unit 1 refueling outage in fall 2010 and the Unit 2 refueling outage in fall 2009. Ultrasonic testing (UT) of the repaired area is also scheduled for the refueling outage in fall 2010. In addition to the visual inspections, the applicant committed to perform supplemental volumetric examinations of liner plate at each unit. A minimum of 75 one-foot square locations will be selected randomly. In addition, a minimum of eight non-random locations will be selected on the basis of perceived greater likelihood of corrosion. Staff agrees that the applicant will examine broad areas for each of the non-random inspections and that they plan to track resolution of any problems identified in any of the inspections. At Unit 1, the non-random UT will begin in the current fuel cycle and are to be completed by December 2010. The random UT will be performed during the next three refueling outages, with all tests to be completed not later than the beginning of period extended operation. At Unit 2, the UT will be completed prior to entering the period of extended operation.

Staff finds that the modified procedures developed following the 2006 event, the additional 100% visual examinations of the liners during the next outages, and the supplemental volumetric examinations to be performed prior to entering the period of extended operation, provide reasonable assurance that the AMP is adequate to manage the aging effects for which it is credited in the license renewal application. The impact of this operating experience on the current operation of the plant is being reviewed and will be resolved under the provisions of the applicant's current 10 CFR Part 50 operating licenses.

We conclude that the proposed inspection programs and related commitments provide reasonable assurance that liner integrity will be adequately maintained during the period of extended operation. Our conclusion is supported by the following observations:

- The mechanism responsible for the through-wall liner penetration in Unit 1 is reasonably well understood. This defect was caused by a wood construction spacer that was not removed as required prior to concrete pour. Wood has the capability to absorb and retain water from the concrete or the atmosphere in the interface between the liner and the concrete. In addition, the testing of the wood revealed that it was acidic and contained 13% water. This acidity could have been the result of boric acid treatment of the wood (a common practice to prevent infestation at the time of construction). This combination of moisture and acidity is corrosive to carbon steel.

- The feature of the supplemental inspection program that addresses this potentially systematic construction error is the non-random UT. FENOC plans to inspect eight broad areas. Some of these will be selected on the basis that additional wood spacers may have been left in the interface between the concrete and the liner during construction. Examinations are intended to ensure that the phenomenon causing the most serious damage is indeed not systematic.

- In addition, 75 or more randomly selected one-foot-square areas will be examined by UT to evaluate the condition of a representative portion of the liner. This examination is intended to determine if unacceptable pitting corrosion is present. The applicant has selected a very stringent failure criterion of >10% localized wall thinning.

- When unit 2 was constructed, welded angle irons were used as spacers between the liner and the first row of re-bar, rather than wood. The absence of wooden spacers significantly reduces the likelihood that the same failure mechanism observed in Unit 1 will occur in Unit 2. Therefore, the supplemental inspection program for Unit 2 on a slower schedule is reasonable.

- The near term 100% visual inspection of all accessible liner surfaces will be valuable in identifying locations for additional UT examinations.

- Based on historical evidence, the opportunity still exists for corrosion caused by the presence of foreign materials. Organic materials such as wood or gloves have been found to be the cause of the same type of damage as that observed in Unit 1 in the containment liners of other plants.

- Inspection of the Unit 1 liner will be completed in time for corrective actions, if required, to be accomplished prior to entering the period of extended operation.

We agree with the staff that there are no issues related to the matters described in 10 CFR 54.29(a)(1) and (a)(2) that preclude renewal of the operating licenses for BVPS, Units 1 and 2. The programs established and committed to by FENOC, including planned supplemental visual and volumetric examinations of the containment liners, provide reasonable assurance that the BVPS, Units 1 and 2,0 can be operated in accordance with its current licensing basis for the period of extended operation without undue risk to the health and safety of the public.

The FENOC application for renewal of the operating licenses for BVPS, Units 1 and 2, should be approved.

Sincerely,

Mario V. Bonaca

Mario V. Bonaca
Chairman

REFERENCES

1. Memorandum dated June 10, 2009 from David Wrona, Office of Nuclear Reactor Regulation, U.S. Nuclear Regulatory Commission, to Edwin M. Hackett, Executive Director, ACRS, transmitting the NRC Final Safety Evaluation Report for Beaver Valley Power Station, Units 1 and 2, License Renewal Application (ML091560099 and ML091600216)

2. Letter dated August 27, 2007, from Peter P. Sena III, FirstEnergy Nuclear Operating Company, to U.S. Nuclear Regulatory Commission, transmitting the Beaver Valley Power Station, Units 1 and 2 License Renewal Application (ML072430913)

3. Safety Evaluation Report Related to the License Renewal of Beaver Valley Power Station, Units 1 and 2, Supplement 1, dated September 2009 (ML092570014)

4. Letter dated July 28, 2009, from Peter P. Sena III, FirstEnergy Nuclear Operating Company, to U.S. Nuclear Regulatory Commission, transmitting Amendment 39, to the Beaver Valley Power Station, Units 1 and 2, License Renewal Application (ML092110117)

5. Letter dated September 2, 2009, from Peter P. Sena III, FirstEnergy Nuclear Operating Company, to U.S. Nuclear Regulatory Commission, transmitting Amendment 40, to the Beaver Valley Power Station, Units 1 and 2, License Renewal Application (ML092510168)

6. Letter dated September 4, 2009, from Peter P. Sena III, FirstEnergy Nuclear Operating Company, to U.S. Nuclear Regulatory Commission, transmitting Amendment 41, to the Beaver Valley Power Station, Units 1 and 2, License Renewal Application (ML092530241)

7. Letter dated September 8, 2009, from Peter P. Sena III, FirstEnergy Nuclear Operating Company, to U.S. Nuclear Regulatory Commission, transmitting Amendment 42, to the Beaver Valley Power Station, Units 1 and 2, License Renewal Application (ML092530242)

8. Letter dated August 27, 2007, from Peter P. Sena III, FirstEnergy Nuclear Operating Company, to U.S. Nuclear Regulatory Commission, transmitting the Beaver Valley Power Station, Units 1 and 2 License Renewal Application (ML072430913)

9. Letter dated July 6, 2009, from John White, U.S. Nuclear Regulatory Commission, to Peter P. Sena III, FirstEnergy Nuclear Operating Company, transmitting the Beaver Valley Power Station License Renewal Inspection Report 05000334/2009006 (ML091870328)

10. Letter dated December 23, 2008, from Richard J. Conte, U.S. Nuclear Regulatory Commission, to Peter P. Sena III, FirstEnergy Nuclear Operating Company, transmitting the Beaver Valley Power Station, License Renewal Inspection Report 05000334/2008007 & 05000412/2008007 (ML083590068)

11.   Letter dated November 5, 2008, from Kent Howard, U.S. Nuclear Regulatory Commission, to Peter P. Sena III, FirstEnergy Nuclear Operating Company, transmitting the Audit Summary Report Regarding the Beaver Valley Power Station, Units 1 and 2, License Renewal Application (ML082140838)

12.   U.S. Nuclear Regulatory Commission, NUREG-1801, Volumes 1 & 2, Revision 1, "Generic Aging Lessons Learned Report," September 2005 (ML052700171).

13.   Letter dated March 13, 2006, from D.E. Graves, Shaw Stone & Webster, Inc. to Robert Dulee, FENOC transmitting the Containment Liner Degradation Report (ML091960569)

14.   Letter dated July 7, 2009, from Theodore S. Robinson, Esquire, Citizen Power to Dr. Dennis C. Bley, ACRS, Regarding the Beaver Valley Power Station Unit 1 Containment Liner (ML091900719)

15.   Letter dated August 27, 2009, from Theodore S. Robinson, Esquire, Citizen Power to Edwin M. Hackett, Executive Director, ACRS, Regarding the Beaver Valley Power Station Unit 1 Containment Liner (ML092460728)

SECTION 6

CONCLUSION

The staff of the United States (US) Nuclear Regulatory Commission (NRC) (the staff) reviewed the license renewal application (LRA) for Beaver Valley Power Station, Units 1 and 2, in accordance with NRC regulations and NUREG-1800, Revision 1, "Standard Review Plan for Review of License Renewal Applications for Nuclear Power Plants," dated September 2005. Title 10, Section 54.29, of the *Code of Federal Regulations* (10 CFR 54.29) sets the standards for issuance of a renewed license.

On the basis of its review of the LRA, the staff concludes that the requirements of 10 CFR 54.29(a) have been met.

The staff noted that any requirements of 10 CFR Part 51, Subpart A, are documented in NUREG-1437, "Generic Environmental Impact Statement for License Renewal of Nuclear Plants (GEIS)" and Supplement 36, "Generic Environmental Impact Statement for License Renewal of Nuclear Plants Regarding Beaver Valley Power Station Units 1 and 2," dated May 14, 2009.

APPENDIX
A

APPENDIX A

BVPS UNITS 1 AND 2 LICENSE RENEWAL COMMITMENTS

During the review of the Beaver Valley Power Station (BVPS), Units 1 and 2, license renewal application (LRA) by the staff of the United States (US) Nuclear Regulatory Commission (NRC) (the staff), FirstEnergy Nuclear Operating Company (the applicant) made commitments related to aging management programs (AMPs) to manage aging effects for structures and components. The following table lists these commitments along with the implementation schedules and sources for each commitment.

BVPS UNIT 1 LICENSE RENEWAL COMMITMENTS

Item Number	Commitment	UFSAR Supplement Section/LRA Section	Enhancement or Implementation Schedule	Source
1	Implement the Buried Piping and Tanks Inspection Program as described in LRA Section B.2.8.	A.1.8 B.2.8	Will be implemented within the 10 years prior to January 29, 2016	LRA
2	Enhance the Closed-Cycle Cooling Water System Program to: • Add the diesel-driven fire pump (Unit 1 only) to the program; • Detail performance testing of heat exchangers and pumps, and provide direction to perform visual inspections of system components; • Identify closed-cycle cooling water system parameters that will be trended to determine if heat exchanger tube fouling or corrosion product buildup exists; • Control performance tests and perform visual inspections at the required frequency.	A.1.9 B.2.9	January 29, 2016	LRA
3	Implement the Electrical Cable Connections Not Subject to 10 CFR 50.49 Environmental Qualification Requirements One-Time Inspection Program as described in LRA Section B.2.10. Prior to implementation of the program, evaluate the program against the final approved version of NRC License Renewal Interim Staff Guidance LR-ISG-2007-02, "Changes To Generic Aging Lesson Learned (GALL) Report Aging Management Program (AMP) XI.E6, "Electrical Cable Connections Not Subject To 10 CFR 50.49 Environmental Qualification Requirements, "" when issued, and revise the program to be consistent with the NRC Interim Staff Guidance.	A.1.10 B.2.10	Will be implemented within the 10 years prior to January 29, 2016	FENOC Letter L-08-262
4	Implement the Electrical Cables and Connections Not Subject to 10 CFR 50.49 Environmental Qualification Requirements Program as described in LRA Section B.2.11.	A.1.11 B.2.11	January 29, 2016	LRA

Item Number	Commitment	UFSAR Supplement Section/LRA Section	Enhancement or Implementation Schedule	Source
5	Implement the Electrical Cables and Connections Not Subject to 10 CFR 50.49 Environmental Qualification Requirements Used in Instrumentation Circuits Program as described in LRA Section B.2.12.	A.1.12 B.2.12	January 29, 2016	LRA
6	Implement the External Surfaces Monitoring Program as described in LRA Section B.2.15.	A.1.15 B.2.15	January 29, 2016 LRA	LRA
7	Enhance the Fire Protection Program to: • Include a new attachment in the BVPS Fire Protection Program administrative procedure to address the Fire Protection Systems that are in scope for license renewal purposes; • Provide details of the NUREG-1801 inspection and testing guidelines, the plant implementation strategy, surveillance test and inspection frequencies (inspection frequency of the Halon and CO_2 systems will be changed to at least once every 6 months), and affected implementing procedure(s); and, • Provide inspection guidance details to include degradation such as concrete cracking and spalling, and loss of material of fire barrier walls, ceilings and floors that may affect the fire rating of the assembly or barrier.	A.1.16 B.2.16	January 29, 2016	LRA and FENOC Letter L-08-375

Item Number	Commitment	UFSAR Supplement Section/LRA Section	Enhancement or Implementation Schedule	Source
8	Enhance the Fire Water System Program to: • Include a program requirement to perform flow test or inspection of all accessible fire water headers and piping during the period of extended operation at an interval determined by the Fire Protection System Engineer; • Include a program requirement that a representative number of fire water piping locations be identified if piping visual inspections are used as an alternative to non-intrusive testing; • Include a program requirement which allows test or inspection results from an accessible Section of pipe to be extrapolated to an inaccessible, but similar Section of pipe. If no similar Section of accessible pipe is available, then alternative testing or inspection activities must be used; • Include a program requirement that, at least once prior to the period of extended operation, all accessible Fire Protection headers and piping shall be flow tested in accordance with NFPA 25 or visually/ultrasonically inspected; • Include steps in the program procedure that require testing or replacement of sprinkler heads that will have been in service for 50 years; and, • Include a program requirement to perform a fire water subsystem internal inspection any time a subsystem (including fire pumps) is breached for repair or maintenance.	A.1.17 B.2.17	Will be implemented within the 10 years prior to January 29, 2016	LRA
9	Enhance the Flux Thimble Tube Inspection Program to: • Include a requirement in the program procedure to state that, if a flux thimble tube cannot be inspected over the tube length (tube length that is subject to wear due to restriction or other defect), and cannot be shown by analysis to be satisfactory for continued service, the thimble tube must be removed from service to ensure the integrity of the Reactor Coolant System pressure boundary.	A.1.19 B.2.19	January 29, 2016	LRA

Item Number	Commitment	UFSAR Supplement Section/LRA Section	Enhancement or Implementation Schedule	Source
10	Enhance the Fuel Oil Chemistry Program to: • Revise the implementing procedure for sampling and testing the diesel-driven fire pump fuel oil storage tank (Unit 1 only) to include a test for particulate and accumulated water in addition to the test for sediment and water; • Generate a new implementing procedure for sampling and testing the security diesel generator fuel oil day tank (Common) for accumulated water, particulate contamination, and sediment/water; and, • Revise implementing procedures to perform UT thickness measurements of accessible above-ground fuel oil tank bottoms at the same frequency as tank cleaning and inspections to ensure that significant degradation is not occurring. For inaccessible tank bottoms, determine tank bottom thickness using an appropriate NDE technique if inspections indicate the presence of significant corrosion.	A.1.20 B.2.20	January 29, 2016	FENOC Letter L-08-316
11	Implement the Inaccessible Medium-Voltage Cables Suitable for Submergence and Not Subject to 10 CFR 50.49 Environmental Qualification Requirements Program as described in LRA Section B.2.21. BVPS commits to implement one of the following prior to entering the period of extended operation: 1. Adopt an acceptable methodology that demonstrates that the in-scope, continuously submerged, inaccessible, medium-voltage cables will continue to perform their intended function during the period of extended operation. -or- 2. Implement measures to minimize cable exposure to significant moisture through dewatering manholes. Incorporate operating experience obtained from dewatering frequency to minimize cable exposure to significant moisture. [Significant moisture is defined as periodic exposures to moisture that last more than a few days (e.g., cable in standing water). Periodic exposures to moisture that last less than a few days (i.e., normal rain and drain) are not significant.] -or- 3. Replace the in-scope, continuously submerged medium-voltage cables with cables designed for submerged service.	A.1.21 B.2.21	January 29, 2016	LRA; FENOC Letter L-08-262; FENOC Letter L-09-057; FENOC Letter L-09-138; and FENOC Letter L-09-151
12	Implement the Inspection of Internal Surfaces in Miscellaneous Piping and Ducting Components Program as described in LRA Section B.2.22.	A.1.22 B.2.22	January 29, 2016	LRA

Item Number	Commitment	UFSAR Supplement Section/LRA Section	Enhancement or Implementation Schedule	Source
13	Enhance the Inspection of Overhead Heavy Load and Light Load (Related to Refueling) Handling Systems Program to: • Include guidance in the program administrative procedure to inspect for loss of material due to corrosion on Unit 1 crane and trolley structural components and rails; and, • Include guidance in the crane and hoist inspection procedures to inspect for loss of material due to corrosion on Unit 1 crane and trolley structural components and rails or extendable arms, as appropriate.	A.1.23 B.2.23	January 29, 2016	LRA
14	Enhance the Masonry Wall Program to: • Include in program scope additional masonry walls identified as having aging effects requiring management for license renewal; and, • Include a requirement in program procedures to incorporate the results of the Masonry Wall Program inspection and document the condition of the walls in the inspection report.	A.1.25 B.2.25	January 29, 2016	FENOC Letter L-08-262
15	Regarding activities for managing the aging of nickel-alloy components and nickel-alloy clad components susceptible to primary water stress corrosion cracking - PWSCC (other than upper reactor vessel closure head nozzles and penetrations), BVPS commits to develop a plant-specific aging management program that will implement applicable: 1. NRC Orders, Bulletins and Generic 2. Letters; and, Staff-accepted industry guidelines.	NONE	January 29, 2016	FENOC Letter L-08-212
16	Implement the One-Time Inspection Program as described in LRA Section B.2.30.	A.1.30 B.2.30	Will be implemented within the 10 years prior to January 29, 2016	LRA
17	Implement the One-Time Inspection of ASME Code Class 1 Small-Bore Piping Program as described in LRA Section B.2.31.	A.1.31 B.2.31	Will be implemented within the 10 years prior to January 29, 2016	LRA

Item Number	Commitment	UFSAR Supplement Section/LRA Section	Enhancement or Implementation Schedule	Source
18	Regarding activities for managing the aging of Reactor Vessel internal components and structures, BVPS commits to: 1. Participate in the industry programs applicable to BVPS Unit 1 for investigating and managing aging effects on reactor internals; 2. Evaluate and implement the results of the industry programs as applicable to the BVPS Unit 1 reactor internals; and, 3. Upon completion of these programs, but not less than 24 months before entering the period of extended operation, submit an inspection plan for the BVPS Unit 1 reactor internals to the NRC for review and approval.	NONE	January 29, 2014	FENOC Letter L-08-212
19	Implement the Selective Leaching of Materials Program as described in LRA Section B.2.36.	A.1.36 B.2.36	January 29, 2016	LRA

Item Number	Commitment	UFSAR Supplement Section/LRA Section	Enhancement or Implementation Schedule	Source
20	Enhance the Structures Monitoring Program to: • Include in program scope additional structures and structural components identified as having aging effects requiring management for license renewal; • Include inspection guidance in program implementing procedures to detect significant cracking in concrete surrounding the anchors of vibrating equipment; • Include a requirement in program procedures to perform opportunistic inspections of normally inaccessible below-grade concrete when excavation work uncovers a significant depth; • Include a requirement in program procedures to perform periodic sampling of groundwater for pH, chloride concentration, and sulfate concentration; and, • Include a requirement in program procedures to monitor elastomeric materials used in seals and sealants, including compressible joints and seals, waterproofing membranes, etc., associated with in-scope structures and structural components for cracking and change in material properties; • Include a requirement in program procedures to perform specific measurements and/or characterizations of structural deficiencies, based on the results of previous inspections and guidance from ACI 349.3R-96, Section 5.1.1, and ACI 201.1 68; • Include a requirement in program procedures to document in the program inspection report a comparison of the results of the program inspections with the results of the previous program inspection; • Include a requirement in program procedures to file the Structures Monitoring Program inspection reports in the BVPS document control system so that inspection results can be more effectively monitored; • Include a requirement in program procedures to apply inspection acceptance criteria based on the results of past inspections and guidance from ACI 349.3R-96, Section 5.1.1. and ACI 201.1-68; and, • Include a requirement in program procedures that noted deficiencies will be reported using the Corrective Action Program.	A.1.39 B.2.39	January 29, 2016 for all enhancements except groundwater sampling (4th bullet). Groundwater sampling will be implemented five (5) years prior to entering the period of extended operation, then continue on a five (5) year interval thereafter.	FENOC Letter L-08-262

Item Number	Commitment	UFSAR Supplement Section/LRA Section	Enhancement or Implementation Schedule	Source
21	With the exception of flexible connections in ventilations systems, prior to the period of extended operation, FENOC will perform repetitive maintenance tasks to replace mechanical system elastomeric components that would otherwise be subject to aging management review. Subsequent frequencies of the repetitive replacements will be based on manufacturer recommendations and applicable operating experience.	NONE	January 29, 2016	FENOC Letter L-08-212
22	Implement the Thermal Aging Embrittlement of Cast Austenitic Stainless Steel (CASS) Program as described in LRA Section B.2.41.	A.1.41 B.2.41	January 29, 2016	LRA
23	Enhance the Water Chemistry Program to: • Change BVPS frequency for reactor coolant silica monitoring to once per week for Operational Modes 1 and 2, and once per day during heatup in Operational Modes 3 and 4 to be consistent with EPRI guidelines.	A.1.42 B.2.42	January 29, 2016	LRA
24	Prior to exceeding the PTS screening criteria for BVPS Unit 1, FENOC will select a flux reduction measure to manage PTS in accordance with the requirements of 10 CFR 50.61. A flux reduction plan submitted for NRC review and approval.	A.2.2.2 4.2.2	A flux reduction plan will be submitted at least 1 year prior to the implementation of the flux reduction measure.	FENOC Letter L-08-124
25	Enhance the Metal Fatigue of the Reactor Coolant Pressure Boundary Program to: • Add a requirement that fatigue will be managed for the NUREG/CR-6260 locations. This requirement will provide that management is accomplished by one or more of the following: 　1. Further refinement of the fatigue analyses to lower the predicted CUFs to less than 1.0; 　2. Management of fatigue at the affected locations by an inspection program that has been reviewed and approved by the NRC (e.g., periodic non-destructive examination of the affected locations at inspection intervals to be determined by a method acceptable to the NRC); or, 　3. Repair or replacement of the affected locations. • Add a requirement that provides for monitoring of the Unit 1 RHR activation transient and establishes an administration limit of 600 cycles for the transient. • Add a requirement to monitor Unit 1 transients where the 60 year projected cycles are used in the environmental fatigue evaluations, and establish an administration limit that is equal to or less than the 60-year projected cycles number.	B.2.27	January 29, 2016	FENOC Letter L-08-209

Item Number	Commitment	UFSAR Supplement Section/LRA Section	Enhancement or Implementation Schedule	Source
26	Evaluate Unit 1 Extended Power Uprate operating experience prior to the period of extended operation for license renewal aging management program adjustments.	Appendix B.2	January 29, 2016	LRA
27	As part of the Reactor Vessel Integrity Program, FENOC will store and maintain Unit 1 standby surveillance capsules in a condition that would permit their future use through the end of the period of extended operation.	B.2.35	Within 30 days following receipt of renewed license	FENOC Letter L-08-143
28	With the exception of underground GeoFlex® fuel oil piping, prior to the period of extended operation, FENOC will perform repetitive maintenance tasks to replace, or to test and replace on condition, mechanical system polymer components that would otherwise be subject to aging management review. Subsequent frequencies of the repetitive tests/replacements will be based on manufacturer recommendations and applicable operating experience.	NONE	January 29, 2016	FENOC Letter L-08-212
29	Confirm the effectiveness of the new license renewal aging management programs based on the incorporation of operating experience by performing a program self assessment of all new license renewal aging management programs. [See NUREG-1800, "Standard Review Plan for Review of License Renewal Applications for Nuclear Power Plants," Appendix A, "Branch Technical Positions," Section A. 1.2.3.10, Items 1 and 2.]	B.2.8 B.2.10 B. 2.11 B. 2.12 B. 2.15 B. 2.21 B. 2.22 B. 2.30 B. 2.31 B. 2.36 B.2.41	January 29, 2021	FENOC Letter L-08-226
30	Enhance the Open-Cycle Cooling Water System Program to: • Include in program scope the Post-Accident Sample System heat exchanger (PAS-E-1) credited with a leakage boundary function; and, • Assess the internal condition of buried piping by opportunistic inspections of header piping internals during removal of expansion joints and inline valves in the headers. Evaluation of inspection results will be documented and trended.	A. 1.32 B. 2.32	January 29, 2016	FENOC Letter L-08-262
31	Implement "needed actions" of MRP-146. These actions include screening, detailed analysis, inspections and temperature monitoring in accordance with the guidelines of MRP-146. FENOC has completed screening of the BVPS RCS branch lines.	NONE	FENOC will perform detailed evaluations (analysis, inspections and/or monitoring) in accordance with MRP-146 schedule requirements, or as established by the MRP committee.	FENOC Letter L-08-287

Item Number	Commitment	UFSAR Supplement Section/LRA Section	Enhancement or Implementation Schedule	Source
32	Supplemental volumetric examinations will be performed on the Unit 1 containment liner prior to the period of extended operation. Seventy five (one foot square) sample locations will be examined. If degradation is identified, the degraded area(s) will be evaluated and follow-up examinations will be performed to ensure the continued reliability of the containment liner.	None	January 29, 2016	FENOC Letter L-090139

BVPS UNIT 2 LICENSE RENEWAL COMMITMENTS

Item Number	Commitment	UFSAR Supplement Section/LRA Section	Enhancement or Implementation Schedule	Source
1	Implement the Buried Piping and Tanks Inspection Program as described in LRA Section B.2.8. Will be implemented within the 10 years prior to May 27, 2027 LRA A.1.8 B.2.8	A.1.8 B.2.8	Will be implemented within the 10 years prior to May 27, 2027	LRA
2	Enhance the Closed-Cycle Cooling Water System Program to: • Add the diesel-driven fire pump (Unit 1 only) and the diesel-driven standby air compressor (Unit 2 only) to the program; • Detail performance testing of heat exchangers and pumps, and provide direction to perform visual inspections of system components; • Identify closed-cycle cooling water system parameters that will be trended to determine if heat exchanger tube fouling or corrosion product buildup exists; • Control performance tests and perform visual inspections at the required frequency.	A.1.9 B.2.9	May 27, 2027	LRA
3	Implement the Electrical Cable Connections Not Subject to 10 CFR 50.49 Environmental Qualification Requirements One-Time Inspection Program as described in LRA Section B.2.10. Prior to implementation of the program, evaluate the program against the final approved version of NRC License Renewal Interim Staff Guidance LR-ISG-2007-02, "Changes To Generic Aging Lesson Learned (GALL) Report Aging Management Program (AMP) XI.E6, "Electrical Cable Connections Not Subject To 10 CFR 50.49 Environmental Qualification Requirements, "" when issued; and revise the program to be consistent with the NRC Interim Staff Guidance.	A.1.10 B.2.10	Will be implemented within the 10 years prior to May 27, 2027	FENOC Letter L-08-262
4	Implement the Electrical Cables and Connections Not Subject to 10 CFR 50.49 Environmental Qualification Requirements Program as described in LRA Section B.2.11.	A.1.11 B.2.11	May 27, 2027	LRA
5	Implement the Electrical Cables and Connections Not Subject to 10 CFR 50.49 Environmental Qualification Requirements Used in Instrumentation Circuits Program as described in LRA Section B.2.12.	A.1.12 B.2.12	May 27, 2027	LRA
6	Implement the Electrical Wooden Poles/Structures Inspection Program as described in LRA Section B.2.13.	A.1.13 B.2.13	Will be implemented within the 5 years prior to May 27, 2027	LRA
7	Implement the External Surfaces Monitoring Program as described in LRA Section B.2.15.	A.1.15 B.2.15	May 27, 2027	LRA

Item Number	Commitment	UFSAR Supplement Section/LRA Section	Enhancement or Implementation Schedule	Source
8	Enhance the Fire Protection Program to: • Include a new attachment to the BVPS Fire Protection Program administrative procedure to address the Fire Protection Systems that are in scope for license renewal purposes; • Provide details of the NUREG-1801 inspection and testing guidelines, the plant implementation strategy, surveillance test and inspection frequencies (inspection frequency of the Halon and CO_2 systems will be changed to at least once every 6 months), and affected implementing procedure(s); and, • Provide inspection guidance details to include degradation such as concrete cracking and spalling, and loss of material of fire barrier walls, ceilings and floors that may affect the fire rating of the assembly or barrier.	A.1.16 B.2.16	May 27, 2027	LRA and FENOC Letter L-08-375
9	Enhance the Fire Water System Program to: • Include a program requirement to perform flow test or inspection of all accessible fire water headers and piping during the period of extended operation at an interval determined by the Fire Protection System Engineer; • Include a program requirement that requires a representative number of fire water piping locations be identified if piping visual inspections are used as an alternative to non-intrusive testing; • Include a program requirement that allows test or inspection results from an accessible Section of pipe to be extrapolated to an inaccessible, but similar Section of pipe. If no similar Section of accessible pipe is available, then alternative testing or inspection activities must be used; • Include a program requirement that, at least once prior to the period of extended operation, all accessible Fire Protection headers and piping shall be flow tested in accordance with NFPA 25 or visually/ultrasonically inspected; • Include steps in the program procedure that require testing or replacement of sprinkler heads that will have been in service for 50 years; and, • Include a program requirement to perform a fire water subsystem internal inspection any time a subsystem (including fire pumps) is breached for repair or maintenance.	A.1.17 B.2.17	Will be implemented within the 10 years prior to May 27, 2027	LRA

Item Number	Commitment	UFSAR Supplement Section/LRA Section	Enhancement or Implementation Schedule	Source
10	Enhance the Flux Thimble Tube Inspection Program to: • Include a requirement in the program procedure to state that, if a flux thimble tube cannot be inspected over the tube length (tube length that is subject to wear due to restriction or other defect), and cannot be shown by analysis to be satisfactory for continued service, the thimble tube must be removed from service to ensure the integrity of the Reactor Coolant System pressure boundary.	A.1.19 B.2.19	May 27, 2027	LRA
11	Enhance the Fuel Oil Chemistry Program to: • Revise the implementing procedure for sampling and testing the diesel-driven fire pump fuel oil storage tank (Unit 1 only) to include a test for particulate and accumulated water in addition to the test for sediment and water; • Generate a new implementing procedure for sampling and testing the security diesel generator fuel oil day tank (Common) for accumulated water, particulate contamination, and sediment/water; and, • Revise implementing procedures to perform UT thickness measurements of accessible above-ground fuel oil tank bottoms at the same frequency as tank cleaning and inspections to ensure that significant degradation is not occurring. For inaccessible tank bottoms, determine tank bottom thickness using an appropriate NDE technique if inspections indicate the presence of significant corrosion.	A.1.20 B.2.20	May 27, 2027	FENOC Letter L-08-316

Item Number	Commitment	UFSAR Supplement Section/LRA Section	Enhancement or Implementation Schedule	Source
12	Implement the Inaccessible Medium- Voltage Cables Suitable for Submergence and Not Subject to 10 CFR 50.49 Environmental Qualification Requirements Program as described in LRA Section B.2.21. BVPS commits to implement one of the following prior to entering the period of extended operation: 1. Adopt an acceptable methodology that demonstrates that the in-scope, continuously submerged, inaccessible, medium-voltage cables will continue to perform their intended function during the period of extended operation. -or- 2. Implement measures to minimize cable exposure to significant moisture through dewatering manholes. Incorporate operating experience obtained from dewatering frequency to minimize cable exposure to significant moisture. [Significant moisture is defined as periodic exposures to moisture that last more than a few days (e.g., cable in standing water). Periodic exposures to moisture that last less than a few days (i.e., normal rain and drain) are not significant.] -or- 3. Replace the in-scope, continuously submerged medium-voltage cables with cables designed for submerged service.	A.1.21 B.2.21	May 27, 2027	LRA; FENOC Letter L-08-262 FENOC Letter L-09-057; FENOC Letter L-09-138; and FENOC Letter L-09-151
13	Implement the Inspection of Internal Surfaces in Miscellaneous Piping and Ducting Components Program as described in LRA Section B.2.22.	A.1.22 B.2.22	May 27, 2027	LRA
14	Enhance the Inspection of Overhead Heavy Load and Light Load (Related to Refueling) Handling Systems Program to: • Include guidance in the program administrative procedure to inspect for loss of material due to corrosion on Unit 2 crane and trolley structural components and rails; and, Include guidance in the crane and hoist inspection procedures to inspect for loss of material due to corrosion on Unit 2 crane and trolley structural components and rails or extendable arms, as appropriate.	A.1.23 B.2.23	May 27, 2027	LRA

Item Number	Commitment	UFSAR Supplement Section/LRA Section	Enhancement or Implementation Schedule	Source
15	Enhance the Masonry Wall Program to: • Include in program scope additional masonry walls identified as having aging effects requiring management for license renewal; and, • Include a requirement in program procedures to incorporate the results of the Masonry Wall Program inspection and document the condition of the walls in the inspection report.	A.1.25 B.2.25	May 27, 2027	FENOC Letter L-08-262
16	Implement the Metal Enclosed Bus Program as described in LRA Section B.2.26.	A.1.26 B.2.26	May 27, 2027	LRA
17	Regarding activities for managing the aging of nickel-alloy components and nickel-alloy clad components susceptible to primary water stress corrosion cracking - PWSCC (other than upper reactor vessel closure head nozzles and penetrations), BVPS commits to develop a plant-specific aging management program that will implement applicable: 1. NRC Orders, Bulletins and Generic Letters; and, 2. Staff-accepted industry guidelines.	NONE	May 27, 2027	FENOC Letter L-08-212
18	Implement the One-Time Inspection Program as described in LRA Section B.2.30.	A.1.30 B.2.30	Will be implemented within the 10 years prior to May 27, 2027	LRA
19	Implement the One-Time Inspection of ASME Code Class 1 Small-Bore Piping Program as described in LRA Section B.2.31.	A.1.31 B.2.31	Will be implemented within the 10 years prior to May 27, 2027	LRA
20	Regarding activities for managing the aging of Reactor Vessel internal components and structures, BVPS commits to: 1. Participate in the industry programs applicable to BVPS Unit 2 for investigating and managing aging effects on reactor internals; 2. Evaluate and implement the results of the industry programs as applicable to the BVPS Unit 2 reactor internals; and, 3. Upon completion of these programs, but not less than 24 months before entering the period of extended operation, submit an inspection plan for the BVPS Unit 2 reactor internals to the NRC for review and approval.	NONE	May 27, 2025	FENOC Letter L-08-212
21	Implement the Selective Leaching of Materials Program as described in LRA Section B.2.36.	A.1.36 B.2.36	May 27, 2027	LRA

Item Number	Commitment	UFSAR Supplement Section/LRA Section	Enhancement or Implementation Schedule	Source
22	Enhance the Structures Monitoring Program to: • Include in program scope additional structures and structural components identified as having aging effects requiring management for license renewal; • Include inspection guidance in program implementing procedures to detect significant cracking in concrete surrounding the anchors of vibrating equipment; • Include a requirement in program procedures to perform opportunistic inspections of normally inaccessible below-grade concrete when excavation work uncovers a significant depth; • Include a requirement in program procedures to perform periodic sampling of groundwater for pH, chloride concentration, and sulfate concentration; • Include a requirement in program procedures to monitor elastomeric materials used in seals and sealants, including compressible joints and seals, waterproofing membranes, etc., associated with in-scope structures and structural components for cracking and change in material properties; • Include a requirement in program procedures to perform specific measurements and/or characterizations of structural deficiencies, based on the results of previous inspections and guidance from ACI 349.3R-96, Section 5.1.1, and ACI 201.1 68; • Include a requirement in program procedures to document in the program inspection report a comparison of the results of the program inspections with the results of the previous program inspection; • Include a requirement in program Procedures to file the Structures Monitoring Program inspection reports in the BVPS document control system so that inspection results can be more effectively monitored; • Include a requirement in program procedures to apply inspection acceptance criteria based on the results of past inspections and guidance from ACI 349.3R-96, Section 5.1.1, and ACI 201.1-68; and, • Include a requirement in program Procedures that noted deficiencies will be reported using the Corrective Action Program.	A.1.39 B.2.39	May 27, 2027 for all enhancements except groundwater sampling (4th bullet). Groundwater sampling will be implemented five (5) years prior to entering the period of extended operation, then continue on a five (5) year interval thereafter.	FENOC Letter L-08-262

Item Number	Commitment	UFSAR Supplement Section/LRA Section	Enhancement or Implementation Schedule	Source
23	With the exception of flexible connections in ventilations systems, prior to the period of extended operation, FENOC will perform repetitive maintenance tasks to replace mechanical system elastomeric components that would otherwise be subject to aging management review. Subsequent frequencies of the repetitive replacements will be based on manufacturer recommendations and applicable operating experience.	NONE	May 27, 2027	FENOC Letter L-08-212
24	Implement the Thermal Aging Embrittlement of Cast Austenitic Stainless Steel (CASS) Program as described in LRA Section B.2.41.	A.1.41 B.2.41	May 27, 2027	LRA
25	Enhance the Water Chemistry Program to: • Change BVPS frequency for reactor coolant silica monitoring to once per week for Operational Modes 1 and 2, and once per day during heatup in Operational Modes 3 and 4, to be consistent with EPRI guidelines.	A.1.42 B.2.42	May 27, 2027	LRA
26	Enhance the Metal Fatigue of the Reactor Coolant Pressure Boundary Program to: • Add a requirement that fatigue will be managed for the NUREG/CR-6260 locations. This requirement will provide that management is accomplished by one or more of the following: 1. Further refinement of the fatigue analyses to lower the predicted CUFs to less than 1.0; 2. Management of fatigue at the affected locations by an inspection program that has been reviewed and approved by the NRC (e.g., periodic non-destructive examination of the affected locations at inspection intervals to be determined by a method acceptable to the NRC); or, 3. Repair or replacement of the affected locations. • Add a requirement that provides for reanalysis, repair, or replacement of the Unit 2 steam generator secondary manway bolts and the steam generator tubes such that the design bases of these components are not exceeded for the period of extended operation. • Add a requirement to monitor Unit 2 transients where the 60-year projected cycles are used in the environmental fatigue evaluations, and establish an administration limit that is equal to or less than the 60-year projected cycles number.	B.2.27	May 27, 2027	FENOC Letter L-08-209

Item Number	Commitment	UFSAR Supplement Section/LRA Section	Enhancement or Implementation Schedule	Source
27	With the exception of underground GeoFlex® fuel oil piping, prior to the period of extended operation, FENOC will perform repetitive maintenance tasks to replace, or to test and replace on condition, mechanical system polymer components that would otherwise be subject to aging management review. Subsequent frequencies of the repetitive tests/replacements will be based on manufacturer recommendations and applicable operating experience.	NONE	May 27, 2027	FENOC Letter L-08-212 and FENOC Letter L-08-376
28	Confirm the effectiveness of the new license renewal aging management programs based on the incorporation of operating experience by performing a program self assessment of all new license renewal aging management programs. [See NUREG-1800, "Standard Review Plan for Review of License Renewal Applications for Nuclear Power Plants," Appendix A, "Branch Technical Positions," Section A. 1.2.3.10, Items 1 and 2.]	B.2.8 B.2.10 B. 2.11 B. 2.12 B. 2.13 B. 2.15 B. 2.21 B. 2.22 B.2.26 B. 2.30 B. 2.31 B. 2.36 B.2.41	May 27, 2032	FENOC Letter L-08-226
29	Evaluate Unit 2 Extended Power Uprate operating experience prior to the period of extended operation for license renewal aging management program adjustments.	Appendix B.2	May 27, 2027	LRA
30	As part of the Reactor Vessel Integrity Program, BVPS will store and maintain Unit 2 standby surveillance capsules in a condition that would permit their future use through the end of the period of extended operation.	B.2.35	Within 30 days following receipt of renewed license	FENOC LetterL-08-143
31	Enhance the Open-Cycle Cooling Water System Program to: • Assess the internal condition of buried piping by opportunistic inspections of header piping internals during removal of expansion joints and inline valves in the headers. Evaluations of inspection results will be documented and trended.	A. 1.32 B.2.32	May 27, 2027	FENOC LetterL-08-262
32	Implement "needed actions" of MRP-146. These actions include screening, detailed analysis, inspections and temperature monitoring in accordance with the guidelines of MRP-146. FENOC has completed screening of the BVPS RCS branch lines.	NONE	FENOC will perform detailed evaluations (analysis, inspections and/or monitoring) in accordance with MRP-146 schedule requirements, or as established by the MRP committee.	FENOC Letter L-08-287

Item Number	Commitment	UFSAR Supplement Section/LRA Section	Enhancement or Implementation Schedule	Source
33	Supplemental volumetric examinations will be performed on the Unit 2 containment liner prior to the period of extended operation. Seventy five (one foot square) sample locations will be examined. If degradation is identified, the degraded area(s) will be evaluated and follow-up examinations will be performed to ensure the continued reliability of the containment liner.	None	May 27, 2027	FENOC Letter L-09-139

APPENDIX
B

APPENDIX B

CHRONOLOGY

This appendix lists chronologically the routine licensing correspondence between the staff of the United States (US) Nuclear Regulatory Commission (NRC) (the staff) and FirstEnergy Nuclear Operating Company (FENOC). This appendix also lists other correspondence on the staff's review of the Beaver Valley Power Station (BVPS), Units 1 and 2 license renewal application (LRA) (under Docket Nos. 50-334 and 50-412).

Date	Accession No.	Subject
March 3, 2003	ML030660587	FENOC Letter No. L-03-035, Beaver Valley Power Station, Unit No. 1 and No. 2, BV-1 Docket No. 50-334, License No. DPR-66, BV-2 Docket No. 50-412, License No. NPF-73, Order Establishing Interim Inspection Requirements for RPV Heads.
March 27, 2003	ML030910021	FENOC Letter No. L-03-053, Beaver Valley Power Station, Unit No. 1 and No. 2, BV-1 Docket No. 50-334, License No. DPR-66, BV-2 Docket No. 50-412, License No. NPF-73, Order (EA-03-009) Relaxation Request.
April 1, 2003	ML030900628	NRC Request for Additional Information, "Beaver Valley Power Station, Unit 1 – Request for Additional Information – Request for Relaxation of Order EA-03-009 (TAC MB8174)"
April 2, 2003	ML030970094	FENOC Letter No. L-03-057, Beaver Valley Power Station, Unit No. 1, BV-1 Docket No. 50-334, License No. DPR-66, Reply to Request for Additional Information Regarding Order EA-03-009 Relaxation Request.
April 7, 2003	ML030970856	NRC Safety Evaluation, "Safety Evaluation for Beaver Valley Unit 1, (Order EA-03-009) Relaxation Request, Examination Coverage for Reactor Pressure Vessel Head Penetration Nozzles (TAC MB8174)."
March 5, 2004	ML040690037	FENOC Letter No. L-04-030, Beaver Valley Power Station, Unit No. 1 and No. 2 BV-1 Docket No. 50-334, License No. DPR-66, BV-2 Docket No. 50-412, License No. NPF-73, Response to First Revised Order (EA-03-009)
April 4, 2004	ML040220181	First Revised NRC Order EA-03-009, "Issuance of First Revised Order (EA-03-009) Establishing Interim Inspection Requirements for Reactor Pressure Vessel Heads at PWRs"
July 10, 2007	ML071800242	Meeting Notice from S. Hoffman to R. Auluck, "FORTHCOMING MEETING WITH FIRSTENERGY NUCLEAR OPERATING COMPANY REGARDING THE BEAVER VALLEY POWER STATION, UNITS 1 AND 2, LICENSE RENEWAL APPLICATION"
July 31, 2007	ML072150044	Letter from P. Sena III to NRC DCD, "Service List Revision for the Beaver Valley Power Station"
August 2, 2007	ML072220388	(PA-LR) Beaver valley Power Station License Renewal Project Presentation to NRC August 2, 2007
August 2, 2007	ML072250023	(PA-LR) Beaver valley Power Station License Renewal Project Presentation to NRC August 2, 2007, Meeting Handouts

Date	Accession No.	Subject
August 27, 2007	ML072410030	Letter from P. Sena III to DCD, "Beaver Valley Power Station, Unit Nos. 1 and 2 BV-1 Docket No. 50-334, License No. DPR-66 BV-2 Docket No. 50-412, License No. NPF-73 Reactor Vessel Capsule Y Report Supplement 1 (Unit 1) and Reactor Vessel Capsule X Report Supplement 1 (Unit 2)"
August 27, 2007	ML072430180	Letter from P. Sena III to DCD, "Beaver Valley Power Station, Unit Nos. 1 and 2 BV-1 Docket No. 50-334, License No. DPR-66 BV-2 Docket No. 50-412, License No. NPF-73 License Renewal Application Boundary Drawings"
August 27, 2007	ML072430182	Beaver Valley Power Station, Unit Nos. 1 and 2 Drawing LR-Structures, Revision 1 "Site Map – In-Scope Structures."
August 27, 2007	ML072430187	Beaver Valley Power Station, Unit Nos. 1 and 2 Drawing 1-00-1, Revision 2 "Drawing Symbol Legend"
August 27, 2007	ML072430189	Beaver Valley Power Station, Unit Nos. 1 and 2 Drawing 1-00-2, Revision 2 "Drawing Symbol Legend"
August 27, 2007	ML072430191	Beaver Valley Power Station, Unit Nos. 1 and 2 Drawing 1-00-3, Revision 0 "System Numbers & System Names"
August 27, 2007	ML072430195	Beaver Valley Power Station, Unit Nos. 1 and 2 Drawing 1-06-1, Revision 5 "Reactor Coolant System (RC)"
August 27, 2007	ML072430196	Beaver Valley Power Station, Unit Nos. 1 and 2 Drawing 1-06-2, Revision 4 "Reactor Coolant System (RC)"
August 27, 2007	ML072430198	Beaver Valley Power Station, Unit Nos. 1 and 2 Drawing 1-06-3, Revision 4 "Reactor Coolant System (RC)"
August 27, 2007	ML072430199	Beaver Valley Power Station, Unit Nos. 1 and 2 Drawing 1-06-4, Revision 3 "Reactor Coolant System (RC)"
August 27, 2007	ML072430201	Beaver Valley Power Station, Unit Nos. 1 and 2 Drawing 1-07-1, Revision 7 "Chemical and Volume Control System (CH)"
August 27, 2007	ML072430204	Beaver Valley Power Station, Unit Nos. 1 and 2 Drawing 1-06-3, Revision 4 "Chemical and Volume Control System (CH)"
August 27, 2007	ML072430214	Beaver Valley Power Station, Unit Nos. 1 and 2 Drawing 1-08-1, Revision 3 "Boron Recovery Degasifiers (BR)"
August 27, 2007	ML072430215	Beaver Valley Power Station, Unit Nos. 1 and 2 Drawing 1-08-2, Revision 3 "Boron Recovery Ion Exchangers (BR)"
August 27, 2007	ML072430216	Beaver Valley Power Station, Unit Nos. 1 and 2 Drawing 1-08-3, Revision 4 "Boron Recovery Evaporator (BR)"
August 27, 2007	ML072430207	Beaver Valley Power Station, Unit Nos. 1 and 2 Drawing 1-07-3, Revision 5 "Chemical and Volume Control System (CH)"
August 27, 2007	ML072430210	Beaver Valley Power Station, Unit Nos. 1 and 2 Drawing 1-07-4, Revision 5 "Chemical and Volume Control System (CH)"
August 27, 2007	ML072430211	Beaver Valley Power Station, Unit Nos. 1 and 2 Drawing 1-07-5, Revision 4 "Chemical and Volume Control System (CH)"
August 27, 2007	ML072430219	Beaver Valley Power Station, Unit Nos. 1 and 2 Drawing 1-08-4, Revision 5 "Boron Recovery Boric Acid Hold Tanks (BR)"
August 27, 2007	ML072430220	Beaver Valley Power Station, Unit Nos. 1 and 2 Drawing 1-08-5, Revision 4 "Boron Recovery PG Water Tanks (BR)"
August 27, 2007	ML072430221	Beaver Valley Power Station, Unit Nos. 1 and 2 Drawing 1-08-6, Revision 4 "Boron Recovery PG Water Tanks (BR)"
August 27, 2007	ML072430223	Beaver Valley Power Station, Unit Nos. 1 and 2 Drawing 1-08-7, Revision 3 "Boron Recovery PG Water Deaerator (BR)"
August 27, 2007	ML072430226	Beaver Valley Power Station, Unit Nos. 1 and 2 Drawing 1-09-1, Revision 5 "Vent and drain System (DV)"
August 27, 2007	ML072430228	Beaver Valley Power Station, Unit Nos. 1 and 2 Drawing 1-09-2, Revision 3 "Vent and drain System (DV)"
August 27, 2007	ML072430231	Beaver Valley Power Station, Unit Nos. 1 and 2 Drawing 1-09-3, Revision 3 "Vent and drain System (DV)"
August 27, 2007	ML072430232	Beaver Valley Power Station, Unit Nos. 1 and 2 Drawing 1-09-4, Revision 3 "Vent and drain System (DV)"

Date	Accession No.	Subject
August 27, 2007	ML072430235	Beaver Valley Power Station, Unit Nos. 1 and 2 Drawing 1-10-1, Revision 5 "Residual Heat Removal System (RH)"
August 27, 2007	ML072430238	Beaver Valley Power Station, Unit Nos. 1 and 2 Drawing 1-11-1, Revision 6 "Safety Injection System (SI)"
August 27, 2007	ML072430239	Beaver Valley Power Station, Unit Nos. 1 and 2 Drawing 1-11-2, Revision 6 "Safety Injection System (SI)"
August 27, 2007	ML072430240	Beaver Valley Power Station, Unit Nos. 1 and 2 Drawing 1-12-1, Revision 4 "Containment Vacuum and Leakage Monitoring System (CV)"
August 27, 2007	ML072430241	Beaver Valley Power Station, Unit Nos. 1 and 2 Drawing 1-13-1, Revision 4 "Containment Depressurization System (CV)"
August 27, 2007	ML072430243	Beaver Valley Power Station, Unit Nos. 1 and 2 Drawing 1-13-2, Revision 3 "Containment Depressurization System (CV)"
August 27, 2007	ML072430244	Beaver Valley Power Station, Unit Nos. 1 and 2 Drawing 1-14A-1, Revision 6 "Sample System (SS)"
August 27, 2007	ML072430245	Beaver Valley Power Station, Unit Nos. 1 and 2 Drawing 1-14A-2, Revision 4 "Sample System (SS)"
August 27, 2007	ML072430247	Beaver Valley Power Station, Unit Nos. 1 and 2 Drawing 1-14A-3, Revision 4 "Sample System (SS)"
August 27, 2007	ML072430248	Beaver Valley Power Station, Unit Nos. 1 and 2 Drawing 1-14C-1, Revision 4 "Post-Accident Sampling System (PAS)"
August 27, 2007	ML072430250	Beaver Valley Power Station, Unit Nos. 1 and 2 Drawing 1-15-1, Revision 4 "Component Cooling Water System (CCR)"
August 27, 2007	ML072430251	Beaver Valley Power Station, Unit Nos. 1 and 2 Drawing 1-15-2, Revision 4 "Component Cooling System (CCR)"
August 27, 2007	ML072430252	Beaver Valley Power Station, Unit Nos. 1 and 2 Drawing 1-15-3, Revision 4 "Component Cooling Water System (CCR)"
August 27, 2007	ML072430253	Beaver Valley Power Station, Unit Nos. 1 and 2 Drawing 1-15-4, Revision 4 "Component Cooling Water System (CCR)"
August 27, 2007	ML072430254	Beaver Valley Power Station, Unit Nos. 1 and 2 Drawing 1-15-5, Revision 5 "Component Cooling Water System (CCR)"
August 27, 2007	ML072430257	Beaver Valley Power Station, Unit Nos. 1 and 2 Drawing 1-16-1, Revision 3 "Ventilation and Air Conditioning – Primary Plant (VS)"
August 27, 2007	ML072430259	Beaver Valley Power Station, Unit Nos. 1 and 2 Drawing 1-17-1, Revision 4 "Liquid Waste Disposal System (LW)"
August 27, 2007	ML072430262	Beaver Valley Power Station, Unit Nos. 1 and 2 Drawing 1-17-2, Revision 4 "Liquid Waste Disposal System (LW)"
August 27, 2007	ML072430263	Beaver Valley Power Station, Unit Nos. 1 and 2 Drawing 1-17-3, Revision 3 "Liquid Waste Disposal System (LW)"
August 27, 2007	ML072430268	Beaver Valley Power Station, Unit Nos. 1 and 2 Drawing 1-18-1, Revision 3 "Solid Waste Disposal System (SW)"
August 27, 2007	ML072430272	Beaver Valley Power Station, Unit Nos. 1 and 2 Drawing 1-18-2, Revision 2 "Solid Waste Disposal System (SW)"
August 27, 2007	ML072430273	Beaver Valley Power Station, Unit Nos. 1 and 2 Drawing 1-18-3, Revision 3 "Solid Waste Disposal System (SW)"
August 27, 2007	ML072430275	Beaver Valley Power Station, Unit Nos. 1 and 2 Drawing 1-19-1, Revision 4 "Gaseous Waste Disposal System (GW)"
August 27, 2007	ML072430276	Beaver Valley Power Station, Unit Nos. 1 and 2 Drawing 1-19-2, Revision 3 "Gaseous Waste Disposal System (GW)"
August 27, 2007	ML072430279	Beaver Valley Power Station, Unit Nos. 1 and 2 Drawing 1-20-1, Revision 3 "Fuel Pool Cooling and Purification System (PC)"
August 27, 2007	ML072430281	Beaver Valley Power Station, Unit Nos. 1 and 2 Drawing 1-21-1, Revision 5 "Main Steam System (MS)"
August 27, 2007	ML072430288	Beaver Valley Power Station, Unit Nos. 1 and 2 Drawing 1-21-2, Revision 3 "Main Steam System (MS)"

Date	Accession No.	Subject
August 27, 2007	ML072430291	Beaver Valley Power Station, Unit Nos. 1 and 2 Drawing 1-21-3, Revision 4 "Main Steam System (MS)"
August 27, 2007	ML072430294	Beaver Valley Power Station, Unit Nos. 1 and 2 Drawing 1-21-3, Revision 4 "Main Steam System (NG)"
August 27, 2007	ML072430296	Beaver Valley Power Station, Unit Nos. 1 and 2 Drawing 1-22-1, Revision 3 "Condensate System (CN)"
August 27, 2007	ML072430300	Beaver Valley Power Station, Unit Nos. 1 and 2 Drawing 1-24-1, Revision 5 "Feederwater System (FW)"
August 27, 2007	ML072430301	Beaver Valley Power Station, Unit Nos. 1 and 2 Drawing 1-24-2, Revision 4 "Feederwater System (FW)"
August 27, 2007	ML072430305	Beaver Valley Power Station, Unit Nos. 1 and 2 Drawing 1-24-3, Revision 3 "Feederwater System (FW)"
August 27, 2007	ML072430307	Beaver Valley Power Station, Unit Nos. 1 and 2 Drawing 1-25-1, Revision 4 "Steam Generator Blowdown System (BD)"
August 27, 2007	ML072430312	Beaver Valley Power Station, Unit Nos. 1 and 2 Drawing 1-26-4, Revision 4 "Steam Generator Blowdown System (BD)"
August 27, 2007	ML072430315	Beaver Valley Power Station, Unit Nos. 1 and 2 Drawing 1-26-6, Revision 3 "Steam Generator Blowdown System (BD)"
August 27, 2007	ML072430316	Beaver Valley Power Station, Unit Nos. 1 and 2 Drawing 1-27-1, Revision 3 "Steam Generator Blowdown System (BD)"
August 27, 2007	ML072430321	Beaver Valley Power Station, Unit Nos. 1 and 2 Drawing 1-27-2, Revision 4 "Auxiliary Steam System (AS)"
August 27, 2007	ML072430322	Beaver Valley Power Station, Unit Nos. 1 and 2 Drawing 1-27-4, Revision 2 "Auxiliary Steam System (AS)"
August 27, 2007	ML072430323	Beaver Valley Power Station, Unit Nos. 1 and 2 Drawing 1-29-1, Revision 3 "Air Conditioning Chilled Water System (AC)"
August 27, 2007	ML072430325	Beaver Valley Power Station, Unit Nos. 1 and 2 Drawing 1-29-2, Revision 3 "Air Conditioning Chilled Water System (AC)"
August 27, 2007	ML072430327	Beaver Valley Power Station, Unit Nos. 1 and 2 Drawing 1-30-1, Revision 4 "River Water System (RW)"
August 27, 2007	ML072430328	Beaver Valley Power Station, Unit Nos. 1 and 2 Drawing 1-30-2, Revision 4 "River Water System (RC)"
August 27, 2007	ML072430338	Beaver Valley Power Station, Unit Nos. 1 and 2 Drawing 1-32-6, Revision 3 "Water Treating System (WT)"
August 27, 2007	ML072430339	Beaver Valley Power Station, Unit Nos. 1 and 2 Drawing 1-32-7, Revision 3 "Water Treating System (WT)"
August 27, 2007	ML072430342	Beaver Valley Power Station, Unit Nos. 1 and 2 Drawing 1-32-8, Revision 3 "Chemical Feed System (WT)"
August 27, 2007	ML072430344	Beaver Valley Power Station, Unit Nos. 1 and 2 Drawing 1-32-9, Revision 5 "Filtered Water System (WF)"
August 27, 2007	ML072430346	Beaver Valley Power Station, Unit Nos. 1 and 2 Drawing 1-33-1, Revision 4 "Fire Protection System (WF)"
August 27, 2007	ML072430352	Beaver Valley Power Station, Unit Nos. 1 and 2 Drawing 1-33-2, Revision 3 "Fire Protection System (WF)"
August 27, 2007	ML072430357	Beaver Valley Power Station, Unit Nos. 1 and 2 Drawing 1-33-3, Revision 4 "Fire Protection –C02 System (FP)"
August 27, 2007	ML072430358	Beaver Valley Power Station, Unit Nos. 1 and 2 Drawing 1-33-4, Revision 4 "Fire Protection – Halon and C02 System (FP)"
August 27, 2007	ML072430330	Beaver Valley Power Station, Unit Nos. 1 and 2 Drawing 1-30-3, Revision 4 "River Water System (RC)"
August 27, 2007	ML072430332	Beaver Valley Power Station, Unit Nos. 1 and 2 Drawing 1-30-4, Revision 4 "River Water System (RC)"
August 27, 2007	ML072430335	Beaver Valley Power Station, Unit Nos. 1 and 2 Drawing 1-32-2, Revision 3 "Water Treating System (WT)"
August 27, 2007	ML072430359	Beaver Valley Power Station, Unit Nos. 1 and 2 Drawing 1-33-7, Revision 3 "Fire Protection System Details (FP)"

Date	Accession No.	Subject
August 27, 2007	ML072430361	Beaver Valley Power Station, Unit Nos. 1 and 2 Drawing 1-33-8, Revision 3 "Fire Protection System Details (FP)"
August 27, 2007	ML072430363	Beaver Valley Power Station, Unit Nos. 1 and 2 Drawing 1-33-3, Revision 4 "Station Compressed Air System (SA)"
August 27, 2007	ML072430370	Beaver Valley Power Station, Unit Nos. 1 and 2 Drawing 1-34-8, Revision 4 "Air System – Intake Structure Watertight Doors (VS)"
August 27, 2007	ML072430371	Beaver Valley Power Station, Unit Nos. 1 and 2 Drawing 1-36-1, Revision 4 "Emergency Diesel Generator Air Start System (DA)"
August 27, 2007	ML072430373	Beaver Valley Power Station, Unit Nos. 1 and 2 Drawing 1-36-2, Revision 4 "Emergency Diesel Generator Fuel Oil System (FO)"
August 27, 2007	ML072430364	Beaver Valley Power Station, Unit Nos. 1 and 2 Drawing 1-34-2, Revision 4 "Instrument Air and Containment Instrument Air System (IA)"
August 27, 2007	ML072430366	Beaver Valley Power Station, Unit Nos. 1 and 2 Drawing 1-34-5, Revision 2 "Instrument Air for PAB (IA)"
August 27, 2007	ML072430368	Beaver Valley Power Station, Unit Nos. 1 and 2 Drawing 1-34-6, Revision 4 "Instrument Air System (IA)"
August 27, 2007	ML072430375	Beaver Valley Power Station, Unit Nos. 1 and 2 Drawing 1-36-3, Revision 3 "Emergency Diesel Generator Lube Oil System (DLO)"
August 27, 2007	ML072430376	Beaver Valley Power Station, Unit Nos. 1 and 2 Drawing 1-36-4, Revision 4 "Emergency Diesel Generator Water Cooling System (DCW)"
August 27, 2007	ML072430379	Beaver Valley Power Station, Unit Nos. 1 and 2 Drawing 1-36-5, Revision 3 "Emergency Diesel Generator - Air Intake and Exchange System (EE)"
August 27, 2007	ML072430380	Beaver Valley Power Station, Unit Nos. 1 and 2 Drawing 1-41A-1, Revision 3 "Hot Water Heating System (HS)"
August 27, 2007	ML072430383	Beaver Valley Power Station, Unit Nos. 1 and 2 Drawing 1-41A-2, Revision 3 "Hot Water Heating System (HS)"
August 27, 2007	ML072430385	Beaver Valley Power Station, Unit Nos. 1 and 2 Drawing 1-41A-3, Revision 4 "Hot Water Heating System (HS)"
August 27, 2007	ML072430387	Beaver Valley Power Station, Unit Nos. 1 and 2 Drawing 1-41B-1, Revision 3 "Glycol Solution Heating System (HS)"
August 27, 2007	ML072430391	Beaver Valley Power Station, Unit Nos. 1 and 2 Drawing 1-41C-1, Revision 4 "Domestic Water System (PL)"
August 27, 2007	ML072430393	Beaver Valley Power Station, Unit Nos. 1 and 2 Drawing 1-41D-1, Revision 3 "Turbine & Service Building & Yard Drains System (RD)"
August 27, 2007	ML072430394	Beaver Valley Power Station, Unit Nos. 1 and 2 Drawing 1-41D-2, Revision 3 "Turbine & Service Building & Yard Drains System (RD)"
August 27, 2007	ML072430395	Beaver Valley Power Station, Unit Nos. 1 and 2 Drawing 1-43-2, Revision 3 "Radiation Monitoring System (RM)"
August 27, 2007	ML072430397	Beaver Valley Power Station, Unit Nos. 1 and 2 Drawing 1-43-3, Revision 3 "Radiation Monitoring System (RM)"
August 27, 2007	ML072430403	Beaver Valley Power Station, Unit Nos. 1 and 2 Drawing 1-43-5, Revision 3 "Radiation Monitoring System (RM)"
August 27, 2007	ML072430407	Beaver Valley Power Station, Unit Nos. 1 and 2 Drawing 1-44A-1, Revision 3 "Control Room Emergency Pressurization Air System (VS)"
August 27, 2007	ML072430410	Beaver Valley Power Station, Unit Nos. 1 and 2 Drawing 1-44A-2, Revision 4 "Control Room Air Conditioning System (VS)"
August 27, 2007	ML072430411	Beaver Valley Power Station, Unit Nos. 1 and 2 Drawing 1-44A-4, Revision 4 "Control Room Air Conditioning System (VS)"

Date	Accession No.	Subject
August 27, 2007	ML0724304132	Beaver Valley Power Station, Unit Nos. 1 and 2 Drawing 1-44B-1, Revision 3 "Air Vent and Cooling System (VS)"
August 27, 2007	ML072430416	Beaver Valley Power Station, Unit Nos. 1 and 2 Drawing 1-44E-1, Revision 3 "Switchgear Air Conditioning System (VS)"
August 27, 2007	ML072430418	Beaver Valley Power Station, Unit Nos. 1 and 2 Drawing 1-44E-3, Revision 4 "Switchgear Air Conditioning System (VS)"
August 27, 2007	ML072430423	Beaver Valley Power Station, Unit Nos. 1 and 2 Drawing 1-44F-1, Revision 3 "Alternate Intake Structure Ventilation System (VS)"
August 27, 2007	ML072430426	Beaver Valley Power Station, Unit Nos. 1 and 2 Drawing 1-45F-1, Revision 4 "Security Diesel Generator System (NHS)"
August 27, 2007	ML072430428	Beaver Valley Power Station, Unit Nos. 1 and 2 Drawing 1-46-1, Revision 4 "Post DBA Hydrogen Control System (HY)"
August 27, 2007	ML072430429	Beaver Valley Power Station, Unit Nos. 1 and 2 Drawing 1-46-2, Revision 4 "Post DBA Hydrogen Analyzer System (HY)"
August 27, 2007	ML072430430	Beaver Valley Power Station, Unit Nos. 1 and 2 Drawing 1-47-1, Revision 3 "Containment System (VS)"
August 27, 2007	ML072430435	Beaver Valley Power Station, Unit Nos. 1 and 2 Drawing 1-58E-1, Revision 4 "ERF Diesel Fuel Oil System (RGF)"
August 27, 2007	ML072430439	Beaver Valley Power Station, Unit Nos. 1 and 2 Drawing 1-58E-2, Revision 4 "ERF Diesel Water System (RGW)"
August 27, 2007	ML072430440	Beaver Valley Power Station, Unit Nos. 1 and 2 Drawing 1-58E-3, Revision 3 "ERF Diesel Lube Oil System (RGO)"
August 27, 2007	ML072430440	Beaver Valley Power Station, Unit Nos. 1 and 2 Drawing 1-58E-3, Revision 3 "ERF Diesel Lube Oil System (RGO)"
August 27, 2007	ML072430481	Beaver Valley Power Station, Unit Nos. 1 and 2 Drawing 2-33-3, Revision 4 "Fire Protection System – Halon Control Building (FPG)"
August 27, 2007	ML072430625	Beaver Valley Power Station, Unit Nos. 1 and 2 Drawing 2-43-18, Revision 3 "Radiation Monitoring System (HVS)"
August 27, 2007	ML072430648	Beaver Valley Power Station, Unit Nos. 1 and 2 Drawing 2-44F-1, Revision 4 "Main & Alternate Intake Structure & Cooling Tower Pump House Ventilation System (HVW)"
August 27, 2007	ML072430914	Letter from P. Sena III to NRC DCD, "Beaver Valley Power Station, Unit Nos. 1 and 2 BV-1 Docket No. 50-334, License No. DPR-66 BV-2 Docket No. 50-412, License No. NPF-73 License Renewal Application Cove r Letter"
August 27, 2007	ML072430916	License-Application for Facility Operating License (Amend/Renewal) DKT 50 Beaver Valley Power Station License Renewal Application. Cover to table 3.3.2-14
August 27, 2007	ML072470493	License-Application for Facility Operating License (Amend/Renewal) DKT 50 Beaver Valley Power Station License Renewal Application. Table 3.3.2-15 to Appendix D
August 27, 2007	ML072470523	License-Application for Facility Operating License (Amend/Renewal) DKT 50 Beaver Valley Power Station License Renewal Application. Appendix E: Applicant's Environmental Report
September 7, 2007	ML072500082	Press Release 07-116 - LICENSE LENEWAL APPLICATION FOR BEAVER VALLEY NUCLEAR PLANT AVAILABLE FOR PUBLIC INSPECTION
September 18, 2007	ML072670501	BV EPU I&C Questions
September 18, 2007	ML072340332	Letter from P. Kuo to P. Sena III, "RECEIPT AND AVAILABILITY OF THE LICENSE RENEWAL APPLICATION FOR THE BEAVER VALLEY POWER STATION, UNITS 1 AND 2"

Date	Accession No.	Subject
September 18, 2007	ML072340374	Federal Register Notice, "NOTICE OF RECEIPT AND AVAILABILITY OF APPLICATION FOR RENEWAL OF BEAVER VALLEY POWER STATION, UNITS 1 AND 2 FACILITY OPERATING LICENSE NOS. DPR-66 AND NPF-73 FOR AN ADDITIONAL 20-YEAR PERIOD DOCKET NOS. 50-334 AND 50-412"
September 26, 2007	ML072330337	SUMMARY OF MEETING HELD ON AUGUST 2, 2007, BETWEEN THE U.S. NUCLEAR REGULATORY COMMISSION STAFF AND FIRSTENERGY NUCLEAR OPERATING COMPANY REPRESENTATIVES TO DISCUSS THE BEAVER VALLEY POWER STATION, UNITS 1 AND 2, LICENSE RENEWAL APPLICATION
October 22, 2007	ML072900312	Letter from P. Kuo to P. Sena III, "DETERMINATION OF ACCEPTABILITY AND SUFFICIENCY FOR DOCKETING, PROPOSED REVIEW SCHEDULE, AND OPPORTUNITY FOR A HEARING REGARDING THE APPLICATION FROM FIRSTENERGY NUCLEAR OPERATING COMPANY, FOR RENEWAL OF THE OPERATING LICENSES FOR THE BEAVER VALLEY POWER STATION, UNITS 1 AND 2"
October 22, 2007	ML072900397	Federal Register Notice - UNITED STATES NUCLEAR REGULATORY COMMISSION FIRSTENERGY NUCLEAR OPERATING COMPANY BEAVER VALLEY POWER STATION, UNITS 1 AND 2 NOTICE OF ACCEPTANCE FOR DOCKETING OF THE APPLICATION AND NOTICE OF OPPORTUNITY FOR HEARING REGARDING RENEWAL OF FACILITY OPERATING LICENSE NOS. DPR-66 AND NPF-73 FOR AN ADDITIONAL 20-YEAR PERIOD DOCKET NOS. 50-334 and 50-412
October 26, 2007	ML072990171	Press Release 07-141 - NRC ANNOUNCES OPPORTUNITY TO REQUEST HEARING ON LICENSE RENEWAL APPLICATION FOR BEAVER VALLEY NUCLEAR PLANT
November 8, 2007	ML073100576	Memorandum from K. Howard to R. Franovich, "FORTHCOMING MEETING TO DISCUSS THE SAFETY REVIEW PROCESS OVERVIEW AND ENVIRONMENTAL SCOPING PROCESS FOR BEAVER VALLEY POWER STATION, UNITS 1 AND 2, LICENSE RENEWAL APPLICATION REVIEW"
November 8, 2007	ML073120267	Press Release 1-07-058 - NRC TO DISCUSS PROCESS FOR REVIEW OF LICENSE RENEWAL APPLICATION FOR BEAVER VALLEY NUCLEAR PLANT, SEEK ENVIRONMENTAL INPUT
December 8, 2007	ML073610255	Letter from P. Sena III to NRC DCD, "Corrections to the Beaver Valley Power Station License Renewal Application Boundary Drawings"
December 21, 2007	ML073610289	Beaver Valley Power Station, Unit Nos. 1 and 2 Drawing 1-15-3, Revision 5 "Component Cooling Water System (CCR)"
December 21, 2007	ML073610290	Beaver Valley Power Station, Unit Nos. 1 and 2 Drawing 1-17-3, Revision 4 "Liquid Waste Disposal System (LW)"
December 21, 2007	ML073610291	Beaver Valley Power Station, Unit Nos. 1 and 2 Drawing 1-30-3, Revision 5 "River Water System (RW)"
December 21, 2007	ML073610294	Beaver Valley Power Station, Unit Nos. 1 and 2 Drawing 1-33-1, Revision 5 "Fire Protection – Water System (FP)"
December 21, 2007	ML073610295	Beaver Valley Power Station, Unit Nos. 1 and 2 Drawing 1-36-2, Revision 5 "Emergency Diesel Generator Fuel Oil System (FO)"
December 21, 2007	ML073610297	Beaver Valley Power Station, Unit Nos. 1 and 2 Drawing 2-36-3, Revision 4 "Diesel Starting Air System (EGA)"

Date	Accession No.	Subject
February 12, 2008	ML080460505	Letter from P. Sena III to NRC DCD, "License Renewal Application Amendment 1: Revision to Reactor Vessel Integrity Aging Management Program Information and Details of Reactor Vessel Surveillance Capsule Withdrawal Schedule Information"
March 3, 2008	ML080590262	Letter from K. Howard to P. Sena III, "REQUEST FOR ADDITIONAL INFORMATION FOR THE REVIEW OF THE BEAVER VALLEY POWER STATION, UNITS 1 AND 2, LICENSE RENEWAL APPLICATION (TAC NOS. MD6593 AND MD6594)"
March 5, 2008	ML080640216	Letter from K. Howard to P. Sena III, "REQUEST FOR ADDITIONAL INFORMATION FOR THE REVIEW OF THE BEAVER VALLEY POWER STATION, UNITS 1 AND 2, LICENSE RENEWAL APPLICATION (TAC NO. MD6593, MD6594)"
March 5, 2008	ML080601020	Letter from K. Howard to P. Sena III, "REQUEST FOR ADDITIONAL INFORMATION FOR THE REVIEW OF THE BEAVER VALLEY POWER STATION, UNITS 1 AND 2, LICENSE RENEWAL APPLICATION (TAC NOS. MD6593 AND MD6594)"
March 21, 2008	ML080720288	Letter from K. Howard to P. Sena III, "REQUEST FOR ADDITIONAL INFORMATION FOR THE REVIEW OF THE BEAVER VALLEY POWER STATION, UNITS 1 AND 2, LICENSE RENEWAL APPLICATION (TAC NOS. MD6593 AND MD6594)"
March 26, 2008	ML080790744	Letter from K. Howard to P. Sena III, "REQUEST FOR ADDITIONAL INFORMATION FOR THE REVIEW OF THE BEAVER VALLEY POWER STATION, UNITS 1 AND 2, LICENSE RENEWAL APPLICATION (TAC NOS. MD6593 AND 6594)"
March 31, 2008	ML080940397	Beaver Valley Power Station, Unit Nos. 1 and 2 Drawing 1-44B-1, Revision 4 "Air Vent and Cooling System (VS)"
March 31, 2008	ML080940399	Beaver Valley Power Station, Unit Nos. 1 and 2 Drawing 2-47-1, Revision 5 "Containment Air Locks & Fuel Transfer Tube System (PHS)"
April 1, 2008	ML080790392	Letter from K. Howard to P. Sena III, "REQUEST FOR ADDITIONAL INFORMATION FOR THE REVIEW OF THE BEAVER VALLEY POWER STATION, UNITS 1 AND 2, LICENSE RENEWAL APPLICATION (TAC NOS. MD6593 AND 6594)"
April 1, 2008	ML080790538	Letter from K. Howard to P. Sena III, "REQUEST FOR ADDITIONAL INFORMATION FOR THE REVIEW OF THE BEAVER VALLEY POWER STATION, UNITS 1 AND 2, LICENSE RENEWAL APPLICATION (TAC NOS. MD6593 AND MD6594)"
April 2, 2008	ML080980212	Letter from P. Sena III to NRC DCD, "Reply to Request for Additional Information for Review of the Beaver Valley Power Station, Units 1 and 2, License Renewal Application (TAC Nos. MD6593 and MD6594), and License Renewal Application Amendment No. 5, L-08-124"
April 3, 2008	ML080790735	Letter from K. Howard to P. Sena III, "REQUEST FOR ADDITIONAL INFORMATION FOR THE REVIEW OF THE BEAVER VALLEY POWER STATION, UNITS 1 AND 2, LICENSE RENEWAL APPLICATION (TAC NOS. MD6593 AND MD6594)"

Date	Accession No.	Subject
April 3, 2008	ML081000296	Letter from P. Sena III to NRC DCD, "Reply to Request for Additional Information for the Review of Beaver Valley Power Station, Units 1 and 2, License Renewal Application (TAC Nos. MD6593 and MD6594), License Renewal Application Amendment No. 4, and Revised License Renewal Boundary Drawings, L-08-123"
April 3, 2008	ML081000322	Beaver Valley Power Station, Unit Nos. 1 and 2 Drawing 1-017-2, Revision 4 "Chemical Volume and Control System"
April 3, 2008	ML081000323	Beaver Valley Power Station, Unit Nos. 1 and 2 Drawing 1-07-3, Revision 6 "Chemical Volume and Control System"
April 3, 2008	ML081000324	Beaver Valley Power Station, Unit Nos. 1 and 2 Drawing 1-08-1, Revision 4 "Boron Recovery Degasifiers"
April 3, 2008	ML081000326	Beaver Valley Power Station, Unit Nos. 1 and 2 Drawing 1-09-2, Revision 4 "Vent and Drain System"
April 3, 2008	ML081000327	Beaver Valley Power Station, Unit Nos. 1 and 2 Drawing 1-12-1, Revision 5 "Containment Vacuum and Leakage Monitoring System"
April 3, 2008	ML081000328	Beaver Valley Power Station, Unit Nos. 1 and 2 Drawing 1-14A-1, Revision 7 "Sample System"
April 3, 2008	ML081000329	Beaver Valley Power Station, Unit Nos. 1 and 2 Drawing LR 1-14A-2, Revision 5 "Sample System"
April 3, 2008	ML081000330	Beaver Valley Power Station, Unit Nos. 1 and 2 Drawing LR 1-19-1, Revision 5 "Gaseous Waste Disposal System"
April 3, 2008	ML081000331	Beaver Valley Power Station, Unit Nos. 1 and 2 Drawing LR 1-24-1, Revision 6 "Feedwater System"
April 3, 2008	ML081000332	Beaver Valley Power Station, Unit Nos. 1 and 2 Drawing LR 1-30-1, Revision 5 "River Water System"
April 3, 2008	ML081000333	Beaver Valley Power Station, Unit Nos. 1 and 2 Drawing LR 1-46-2, Revision 5 "Post DBA Hydrogen System"
April 17, 2008	ML080980483	Letter from K. Howard to P. Sena III, "REQUEST FOR ADDITIONAL INFORMATION FOR THE REVIEW OF THE BEAVER VALLEY POWER STATION, UNITS 1 AND 2, LICENSE RENEWAL APPLICATION (TAC NOS. MD6593 AND MD6594)"
April 17, 2008	ML081050333	Letter from K. Howard to P. Sena III, "REQUEST FOR ADDITIONAL INFORMATION FOR THE REVIEW OF THE BEAVER VALLEY POWER STATION, UNITS 1 AND 2, LICENSE RENEWAL APPLICATION (TAC NOS. MD6593 AND MD6594)"
April 18, 2008	ML081130155	Letter from P. Sena III to NRC DCD, "Reply to Request for Additional Information for Review of the Beaver Valley Power Station, Units 1 and 2, License Renewal Application (TAC Nos. MD6593 and MD6594), and License Renewal Application Amendment No. 6, L-08-143"
April 25, 2008	ML081200596	Letter from P. Sena III to NRC DCD, "Request for Schedule Change for Advisory Committee on Reactor Safeguards Subcommittee Review of the Beaver Valley Power Station, Units 1 and 2, License Renewal Safety Evaluation Report, L-08-145"
April 25, 2008	ML081200597	Letter from P. Sena III to NRC DCD, "Reply to Request for Additional Information for the Review of the Beaver Valley Power Station, Units 1 and 2. License Renewal Application (TAC Nos. MD6593 and MD6594), L-08-144"
April 28, 2008	ML081130394	Letter from K. Howard to P. Sena III, "REQUEST FOR ADDITIONAL INFORMATION FOR THE REVIEW OF THE BEAVER VALLEY POWER STATION, UNITS 1 AND 2, LICENSE RENEWAL APPLICATION (TAC NOS. MD6593 AND MD6594)"

Date	Accession No.	Subject
April 30, 2008	ML081050270	Letter from K. Howard to P. Sena III, "REQUEST FOR ADDITIONAL INFORMATION FOR THE REVIEW OF THE BEAVER VALLEY POWER STATION, UNITS 1 AND 2, LICENSE RENEWAL APPLICATION (TAC NOS. MD6593 AND MD6594)"
April 30, 2008	ML081230618	Letter from P. Sena III to NRC DCD, "Reply to Request for Additional Information for Review of the Beaver Valley Power Station, Units 1 and 2, License Renewal Application (TAC Nos. MD6593 and MD6594), and License Renewal Application Amendment No. 7, L-08-146"
May 2, 2008	ML081270236	Letter from P. Sena III to NRC DCD, "Reply to Request for Additional Information for Review of the Beaver Valley Power Station, Units 1 and 2, License Renewal Application (TAC Nos. MD6593 and MD6594), and License Renewal Application Amendment No. 7, L-08-174"
May 5, 2008	ML081280490	Letter from P. Sena III to NRC DCD, "Reply to Request for Additional Information for Review of the Beaver Valley Power Station, Units 1 and 2, License Renewal Application (TAC Nos. MD6593 and MD6594), and License Renewal Application Amendment No. 7, L-08-149"
May 5, 2008	ML081410416	Letter from P. Sena III to NRC DCD, "Correction to Reply to Request for Additional Information for the Review of Beaver Valley Power Station, Units 1 and 2, License Renewal Application (TAC Nos. MD6593 and MD6594), License Renewal Application Amendment No. 8, and Revised License Renewal Boundary Drawings, L-08-150"
May 8, 2008	ML081050543	Letter from K. Howard to P. Sena III, "REQUEST FOR ADDITIONAL INFORMATION FOR THE REVIEW OF THE BEAVER VALLEY POWER STATION, UNITS 1 AND 2, LICENSE RENEWAL APPLICATION (TAC NOS. MD6593 AND MD6594)"
May 8, 2008	ML081160467	Letter from K. Howard to P. Sena III, "REQUEST FOR ADDITIONAL INFORMATION FOR THE REVIEW OF THE BEAVER VALLEY POWER STATION, UNITS 1 AND 2, LICENSE RENEWAL APPLICATION (TAC NOS. MD6593 AND MD6594)"
May 8, 2008	ML081200920	Letter from K. Howard to P. Sena III, "REQUEST FOR ADDITIONAL INFORMATION FOR THE REVIEW OF THE BEAVER VALLEY POWER STATION, UNITS 1 AND 2, LICENSE RENEWAL APPLICATION (TAC NOS. MD6593 AND MD6594)"
May 8, 2008	ML081410417	Beaver Valley Power Station, Unit Nos. 1 and 2 Drawing LR 2-30-1, Revision 4 "Service Water System (SWS)"
May 8, 2008	ML081410419	Beaver Valley Power Station, Unit Nos. 1 and 2 Drawing LR – Structures, Revision 2, site Map – In-Scope Structures.
May 15, 2008	ML081120539	Letter from K. Howard to P. Sena III, "REQUEST FOR ADDITIONAL INFORMATION FOR THE REVIEW OF THE BEAVER VALLEY POWER STATION, UNITS 1 AND 2, LICENSE RENEWAL APPLICATION (TAC NOS. MD6593 AND MD6594)"
May 19, 2008	ML081420368	Letter from P. Sena III to NRC DCD, "Reply to Request for Additional Information for the Review of the Beaver Valley Power Station, Units 1 and 2, License Renewal Application (TAC Nos. MD6593 and MD6594), L-08-169"

Date	Accession No.	Subject
May 19, 2008	ML081440543	Letter from P. Sena III to NRC DCD, "Reply to Request for Additional Information for the Review of the Beaver Valley Power Station, Units 1 and 2, License Renewal Application (TAC Nos. MD6593 and MD6594), License Renewal Application Amendment No. 9. and Revised License Renewal Boundary Drawings, L-08-170"
May 19, 2008	ML081440711	Beaver Valley Power Station, Unit Nos. 1 and 2 Drawing 2-34-2, Revision 2 " Station Instrument Air (IAS)".
May 19, 2008	ML081440712	Beaver Valley Power Station, Unit Nos. 1 and 2 Drawing 2-34-3, Revision 7 " Containment Instrument Air System (IAC)".
May 19, 2008	ML081440714	Beaver Valley Power Station, Unit Nos. 1 and 2 Drawing 2-34-10, Revision 1 "Containment Instrument Air (IAC)".
May 19, 2008	ML081440716	Beaver Valley Power Station, Unit Nos. 1 and 2 Drawing 2-34-11, Revision 2 "Instrument Air Standby Train (IAS)".
May 22, 2008	ML081130412	Letter from K. Howard to P. Sena III, "REQUEST FOR ADDITIONAL INFORMATION FOR THE REVIEW OF THE BEAVER VALLEY POWER STATION, UNITS 1 AND 2, LICENSE RENEWAL APPLICATION (TAC NOS. MD6593 AND MD6594)"
May 22, 2008	ML081360557	Letter from K. Howard to P. Sena III, "REVISION OF SCHEDULE FOR THE CONDUCT OF REVIEW OF THE BEAVER VALLEY POWER STATION, UNITS 1 AND 2, LICENSE RENEWAL APPLICATION (TAC NOS. MD6593, MD6594, MD6595 AND MD6596)"
May 28, 2008	ML081150577	Letter from K. Howard to P. Sena III, "REQUEST FOR ADDITIONAL INFORMATION FOR THE REVIEW OF THE BEAVER VALLEY POWER STATION, UNITS 1 AND 2, LICENSE RENEWAL APPLICATION (TAC NOS. MD6593 AND MD6594)"
May 28, 2008	ML081510406	Letter from Pete Sena III to NRC DCD, "Reply to Request for Additional Information for the Review of the Beaver Valley Power Station, Units 1 and 2, License Renewal Application (TAC Nos. MD6593 and MD6594)" L-08-180
June 2, 2008	ML081560245	Letter from Pete Sena III to NRC DCD, "Reply to Request for Additional Information for the Review of the Beaver Valley Power Station, Units 1 and 2, License Renewal Application (TAC Nos. MD6593 and MD6594)" L-08-147
June 4, 2008	ML081430687	Letter from K. Howard to P. Sena III, "REQUEST FOR ADDITIONAL INFORMATION FOR THE REVIEW OF THE BEAVER VALLEY POWER STATION, UNITS 1 AND 2, LICENSE RENEWAL APPLICATION (TAC NOS. MD6593 AND MD6594)"
June 5, 2008	ML081540195	Letter from R. Franovich to P. Sena III, "REQUEST FOR ADDITIONAL INFORMATION FOR THE REVIEW OF THE BEAVER VALLEY POWER STATION, UNITS 1 AND 2, LICENSE RENEWALAPPLICATION (TAC NOS. MD6593 AND MD6594)"
June 6, 2008	ML081620356	Letter from Pete Sena III to NRC DCD, "Reply to Request for Additional Information for the Review of the Beaver Valley Power Station, Units 1 and 2, License Renewal Application (TAC Nos. MD6593 and MD6594) and License Renewal Application Amendment No. 10" L-08-181
June 9, 2008	ML081640097	Letter from Pete Sena III to NRC DCD, "Reply to Request for Additional Information for the Review of the Beaver Valley Power Station, Units 1 and 2, License Renewal Application (TAC Nos. MD6593 and MD6594) and License Renewal Application Amendment No. 12" L-08-190

Date	Accession No.	Subject
June 9, 2008	ML081640505	Letter from Pete Sena III to NRC DCD, "Reply to Request for Additional Information for the Review of the Beaver Valley Power Station, Units 1 and 2, License Renewal Application (TAC Nos. MD6593 and MD6594), License Renewal Application Amendment No. 11, and Revised License Renewal Boundary Drawings" L-08-189
June 9, 2008	ML081640592	Beaver Valley Power Station, Unit Nos. 1 and 2 Drawing LR 2-30-1, Revision 4 "Sample System (SS)"
June 9, 2008	ML081640593	Beaver Valley Power Station, Unit Nos. 1 and 2 Drawing LR 1-21-1, Revision 6 "Main Steam System (MS)"
June 9, 2008	ML081640594	Beaver Valley Power Station, Unit Nos. 1 and 2 Drawing LR 1-22-1, Revision 4 "Condensate System (CN)"
June 9, 2008	ML081640595	Beaver Valley Power Station, Unit Nos. 1 and 2 Drawing LR 1-24-2, Revision 5 "Feedwater System (FW)"
June 9, 2008	ML081640596	Beaver Valley Power Station, Unit Nos. 1 and 2 Drawing LR 1-24-4, Revision 4 "Feedwater System (FW)"
June 9, 2008	ML081640597	Beaver Valley Power Station, Unit Nos. 1 and 2 Drawing LR 1-25-1, Revision 5 "Steam Generator Blowdown System (BD)"
June 9, 2008	ML081640598	Beaver Valley Power Station, Unit Nos. 1 and 2 Drawing LR 1-32-7, Revision 4 "Water Treating System (WT)"
June 9, 2008	ML081640599	Beaver Valley Power Station, Unit Nos. 1 and 2 Drawing LR 1-45F-1, Revision 5 "Security Diesel Generator System (NHS)"
June 9, 2008	ML081640600	Beaver Valley Power Station, Unit Nos. 1 and 2 Drawing LR 1-58E-1, Revision 5 "Diesel Fuel Oil System (RGF)"
June 9, 2008	ML081640601	Beaver Valley Power Station, Unit Nos. 1 and 2 Drawing LR 2-14A-1, Revision 6 "Reactor Plant Sample System (SSR)"
June 9, 2008	ML081640602	Beaver Valley Power Station, Unit Nos. 1 and 2 Drawing LR 2-15-1, Revision 5 "Primary Component Coolant System (CCP)"
June 9, 2008	ML081640603	Beaver Valley Power Station, Unit Nos. 1 and 2 Drawing LR 2-15-6, Revision 5 "Primary Component Coolant System (CCP)"
June 9, 2008	ML081640604	Beaver Valley Power Station, Unit Nos. 1 and 2 Drawing LR 2-24-2A, Revision 4 "main Feedwater System (FWS)"
June 9, 2008	ML081640605	Beaver Valley Power Station, Unit Nos. 1 and 2 Drawing LR 2-27A-2, Revision 4 "Auxiliary Steam and Condensate System (ASS)"
June 16, 2008	ML08700236	Letter from Pete Sena III to NRC DCD, "Reply to Request for Additional Information for the Review of the Beaver Valley Power Station, Units 1 and 2, License Renewal Application (TAC Nos. MD6593 and MD6594) and License Renewal Application Amendment No. 13" L-08-188
June 17, 2008	ML08700652	Letter from Pete Sena III to NRC DCD, "Reply to Request for Additional Information for the Review of the Beaver Valley Power Station, Units 1 and 2, License Renewal Application (TAC Nos. MD6593 and MD6594) and License Renewal Application Amendment No. 14" L-08-191
July 21, 2008	ML082060074	Letter from Pete Sena III to NRC DCD, "Reply to Request for Additional Information for the Review of the Beaver Valley Power Station, Units 1 and 2. License Renewal Application (TAC Nos. MD6593 and MD6594), and License Renewal Application Amendment No. 17, L-08-212"
July 24, 2008	ML082100073	Letter from Pete Sena III to NRC DCD, "Reply to Request for Additional Information for the Review of the Beaver Valley Power Station, Units 1 and 2, License Renewal Application (TAC Nos. MD6593 and MD6594) and License Renewal Application Amendment No. 19, L-08-213"

Date	Accession No.	Subject
July 24, 2008	ML082100075	Letter from Pete Sena III to NRC DCD, "Supplement to Reply to Request for Additional Information for the Review of the Beaver Valley Power Station, Units 1 and 2, License Renewal Application (TAC Nos. MD6593 and MD6594), L-08-227"
July 24, 2008	ML082100307	Letter from Pete Sena III to NRC DCD, "Reply to Request for Additional Information for the Review of the Beaver Valley Power Station, Units 1 and 2, License Renewal Application (TAC Nos. MD6593 and MD6594) and License Renewal Application Amendment No. 18, L-08-211"
July 24, 2008	ML082100375	Beaver Valley Power Station, Unit Nos. 1 and 2 Drawing LR-Structures, Revision 4, Site Map – In-Scope Structures."
August 1, 2008	ML082180124	Letter from Pete Sena III to NRC DCD, "Responses to a Request for Additional Information in Support of License Amendment Request No. 07-005 (TAC Nos. MD7531 and MD7532), L-08-229"
August 7, 2008	ML082120586	SUMMARY OF TELEPHONE CONFERENCE CALL HELD ON JULY 1, 2008, BETWEEN THE U.S. NUCLEAR REGULATORY COMMISSION AND FIRSTENERGY NUCLEAR OPERATING COMPANY, CONCERNING REQUESTS FOR ADDITIONAL INFORMATION PERTAINING TO THE EAVER VALLEY POWER STATION, UNITS 1 AND 2, LICENSE RENEWAL APPLICATION
August 13, 2008	ML082270597	Letter from Pete Sena III to NRC DCD, "Reply to Request for Additional Information for the Review of the Beaver Valley Power Station, Units 1 and 2, License Renewal Application (TAC Nos. MD6593 and MD6594) and License Renewal Application Amendment No. 20, L-08-260"
August 22, 2008	ML082390814	Letter from Pete Sena III to NRC DCD, "Reply to Request for Additional Information for the Review of the Beaver Valley Power Station, Units 1 and 2, License Renewal Application (TAC Nos. MD6593 and MD6594) and License Renewal Application Amendment No. 21, L-08-226"
August 22, 2008	ML082390815	Letter from Pete Sena III to NRC DCD, "Reply to Request for Additional Information for the Review of the Beaver Valley Power Station, Units 1 and 2, License Renewal Application (TAC Nos. MD6593 and MD6594) and License Renewal Application Amendment No. 22, L-08-269"
August 22, 2008	ML082390816	Letter from Pete Sena III to NRC DCD, "Schedule for Submittal of Annual Update for the Beaver Valley Power Station, Units 1 and 2, License Renewal Application (TAC Nos. MD6593 and MD6594), L-08-261"
September 3, 2008	ML082401708	Letter from K. Howard to P. Sena III, "REQUEST FOR ADDITIONAL INFORMATION FOR THE REVIEW OF THE BEAVER VALLEY POWER STATION, UNITS 1 AND 2, LICENSE RENEWAL APPLICATION (TAC NOS. MD6593 AND MD6594)"
September 8, 2008	ML082550686	Beaver Valley Power Station, Unit Nos. 1 and 2 Drawing LR 2-34-2, Revision 3 "Station Instrument Air (IAS)"
September 8, 2008	ML082550687	Beaver Valley Power Station, Unit Nos. 1 and 2 Drawing LR 1-24-2, Revision 6 "Feedwater System (FW)"
September 8, 2008	ML082550688	Beaver Valley Power Station, Unit Nos. 1 and 2 Drawing LR 1-41C-1, Revision 5 "Domestic Water System (PL)"
September 8, 2008	ML082550689	Beaver Valley Power Station, Unit Nos. 1 and 2 Drawing LR 2-31-1, Revision 5 "Demineralized Water System (WTD)"
September 8, 2008	ML082550693	Letter from Pete Sena III to NRC DCD, "License Renewal Application Amendment No. 23 (TAC Nos. MD6593 and MD6594) and Revised License Renewal Boundary Drawings, L-08-262"

Date	Accession No.	Subject
September 11, 2008	ML082730717	Letter from Pete Sena III to NRC DCD, "License Renewal Application Amendment No. 24 (TAC Nos. MD6593 and MD6594), L-08-263"
September 22, 2008	ML082740204	Letter from Pete Sena III to NRC DCD, "Submittal of Corrected WCAP15571 Supplement 1 and WCAP15571-NP, L-08-289"
October 2, 2008	ML082800177	Letter from Pete Sena III to NRC DCD, "Reply to Request for Additional Information for the Review of the Beaver Valley Power Station, Units 1 and 2, License Renewal Application (TAC Nos. MD6593 and MD6594) and License Renewal Application Amendment No. 25, L-08-287"
October 2, 2008	ML082800180	Letter from Pete Sena III to NRC DCD, "License Renewal Application Amendment No. 26 (TAC Nos. MD6593 and MD6594), L-08-316"
October 3, 2008	ML082810100	Letter from Pete Sena III to NRC DCD, "Reply to Request for Additional Information for the Review of the Beaver Valley Power Station, Units 1 and 2, License Renewal Application (TAC Nos. MD6593 and MD6594) and License Renewal Application Amendment No. 27, L-08-310"
October 3, 2008	ML082810106	Letter from Pete Sena III to NRC DCD, "Supplement to Reply to Request for Additional Information for the Review of the Beaver Valley Power Station, Units 1 and 2, License Renewal Application (TAC Nos. MD6593 and MD6594) and License Renewal Application Amendment No. 28, L-08-309"
October 10, 2008	ML082890154	Letter from Pete Sena III to NRC DCD, "Supplemental Information for the Review of the Beaver Valley Power Station, Units 1 and 2, License Renewal Application (TAC Nos. MD6593 and MD6594), L-08-324"
October 10, 2008	ML082900489	Letter from Pete Sena III to NRC DCD, "Response to Request for Additional Information - 2007 Steam Generator Tube Inspections (TAC No. MD8392), L-08-297"
October 24, 2008	ML083030071	Letter from Pete Sena III to NRC DCD, "Supplement to Information Provided in License Renewal Application Amendment No. 23 (TAC Nos. MD6593 and MD6594) Regarding Submersible Cable Suitability, L-08-288"
October 24, 2008	ML083040268	Letter from Pete Sena III to NRC DCD, "License Renewal Application Amendment No. 29 (Annual Update) (TAC Nos. MD6593 and MD6594) and Revised License Renewal Application Boundary Drawing, L-08-292"
October 30, 2008	ML083040266	Beaver Valley Power Station, Unit Nos. 1 and 2 Drawing LR 1-21-1, Revision 7 "Main Steam System (MS)"
November 04, 2008	ML083090886	Press Release I-08-060, "NRC TO DISCUSS RESULTS OF LICENSE RENEWAL INSPECTION FOR BEAVER VALLEY NUCLEAR POWER PLANT ON NOV. 12"
November 11, 2008	ML082140838	Letter from K. Howard to P. Sena III, "AUDIT REPORT REGARDING THE BEAVER VALLEY POWER STATION, UNIT 1 AND 2, LICENSE RENEWAL APPLICATION"
November 13, 2008	ML083010249	Summary of Telephone Conference Call Held on 10/8/08, between the NRC and FirstEnergy Nuclear Operating Company, Concerning Open Items Pertaining to the BVPS, Units 1 and 2, License Renewal SER.
November 13, 2008	ML083020290	Summary of Telephone Conference Call Held on 8/28/08, between the NRC and FENOC, Concerning RAI Pertaining to the BVPS, Units 1 and 2, LRA.
December 15, 2008	ML083250640	Summary of Telephone Conference Call Held 11/14/08, between the NRC and FENOC, Concerning RAIs Pertaining to the Beaver Valley Power Station, Units 1 and 2, License Renewal Application.

Date	Accession No.	Subject
December 17, 2008	ML083230667	10/22/2008 - Summary of Telephone Conference Call Between the NRC and FirstEnergy Nuclear Operating Company, Concerning RAIs Pertaining to the Beaver Valley Power Station, Units 1 and 2, License Renewal Application.
December 19, 2008	ML083250420	Summary of Telephone Conference Call Held on 11/17/08, between the NRC and FENOC, Concerning Requests for Additional Information Pertaining to the Beaver Valley Power Station, Units 1 and 2, LRA.
December 19, 2008	ML083590223	Letter from Pete Sena III to NRC DCD, "Supplemental Information for the Review of the Beaver Valley Power Station, Units 1 and 2, License Renewal Application (TAC Nos. MD6593 and MD6594), and License Renewal Amendment No. 32," L-08-292"
December 22, 2008	ML083650066	Beaver Valley, Units 1 and 2 - License Renewal Application, Amendment No. 33.
January 09, 2009	ML083660029	10/30/08 Summary of Public Meeting on the Draft Supplemental Environmental Impact Statement Regarding the Beaver Valley Power Station, License Renewal Review.
January 09, 2009	ML090080046	Safety Evaluation Report With Open Item Related To The License Renewal Of Beaver Valley Power Station, Units 1 and 2.
January 09, 2009	ML090120360	Beaver Valley, Units 1 & 2, Safety Evaluation Report.
January 19, 2009	ML090220216	Beaver Valley, Units 1 & 2, Supplemental Information for the Review of the License Renewal Application and License Renewal Application Amendment No. 34. (
January 21, 2009	ML083500325	Summary of Telephone Conf Call Held on Sept. 26, 2008 Between NRC and FirstEnergy Nuclear Operating Company Concerning the RAI Pertaining to the Refurbishment Activities at the BVPS. Units 1&2 LRA.
February 04, 2009	ML090500815	Transcript of the ACRS Plant License Renewal Subcommittee Meeting (Beaver Valley) on February 4, 2009, Pages 1-146.
February 24, 2009	ML090220553	Request for Additional Information for the Review of the Beaver Valley Power Station, Units 1 and 2, License Renewal Application. (
March 24, 2009	ML090850433	Beaver Valley, Units 1 & 2, Reply to Request for Additional Information for Review of License Renewal Application.
May 07, 2009	ML091260025	Request for Additional Information for the Review of the Beaver Valley Power Station, Units 1 and 2, License Renewal Application.
May 12, 2009	ML091250413	Final Supplement 36 to the Generic Environmental Impact Statement for License Renewal of Nuclear Plants (GEIS) Regarding Beaver Valley Power Station, Units 1 and 2 (TAC Nos. MD6595 and MD6596).
May 12, 2009	ML091260024	Final Supplement 36 to NUREG-1437, "Generic Environmental Impact Statement for License Renewal of Nuclear Plants (BVPS).

Date	Accession No.	Subject
May 14, 2009	ML091250363	Notice of Availability of the Final Plant-Specific Supplement 36 to the GEIS for License Renewal of Nuclear Plants Regarding Beaver Valley Power Station.
May 14, 2009	ML091380033	Beaver Valley, Unit 1 and 2 - License Renewal Application, Amendment No. 36.
May 20, 2009	ML091400166	Docketing of NRC Teleconference Notes Pertaining to the License Renewal of the Beaver Valley Power Station, Units 1 and 2.
May 20, 2009	ML091420273	Beaver Valley, Units 1 and 2, License Renewal Application Amendment No. 37
May 31, 2009	ML091260011	NUREG-1437 Supplement 36 "Generic Environmental Impact Statement for License Renewal of Nuclear Plants Regarding Beaver Valley Power Station Units 1 and 2" Final Report.
June 01, 2009	ML091540012	Beaver Valley, Units 1 and 2 - Reply to Request for Additional Information on License Renewal Application Amendment No. 38
June 08, 2009	ML091560200	Docketing of NRC Teleconference Notes Pertaining to the License Renewal of the Beaver Valley Power Station, Units 1 and 2
September 16, 2009	ML091980300	Report on the Safety Aspects of the License Renewal Application for the Beaver Valley Power Station, Units 1 and 2

APPENDIX
C

APPENDIX C

PRINCIPAL CONTRIBUTORS

This appendix lists the principal contributors for the development of this safety evaluation report (SER) and their areas of responsibility.

Name	Responsibility
H. Ashar	Structural Engineering
D. Ashley	SER Support
R. Auluck	Management Oversight
J. Bettle	Containment and Ventilation
T. Chan	Management Oversight
K. Chang	Audit Team Member
Y. Chung	Audit Team Member
G. Cranston	Management Oversight
M. Cunningham	Management Oversight
Dr. J. Davis	Audit Team Member
R. Dennig	Management Oversight
K Desai	Reactor Systems Engineer
J. Dozier	Management Oversight
Y. Edmonds	SER Support
M. Evans	Management Oversight
J. Fair	Structural Engineering
F. Farzam	Structural Engineering
S. Figueroa	SER Support
R. Franovich	Management Oversight
Q. Gan	Mechanical Engineering
S. Gardocki	Balance of Plant
R. Green	SER Support
D. Harrison	Management Oversight
M. Hartzman	Structural Engineering
M. Heath	SER Support
B. Heida	Containment and Ventilation
P. Hiland	Management Oversight
A. Hiser	Management Oversight
Dr. D. Hoang	Civil & Structural Engineering
B. Holian	Management Oversight
K. Howard	Lead Project Manager
N. Iqbal	Fire Protection
I. King	SER Support
M. Khanna	Management Oversight
A. Klein	Management Oversight
P.T. Kuo	Management Oversight
S. Lee	Management Oversight
B. Lehman	Civil & Structural Engineering
R. Li	Structural Engineering
L. Lois	Reactor Systems
S. Lopas	Project Manager
K. Manoly	Management Oversight
C. Marks	Consultant
J. Medoff	Audit Team Member

Name	Responsibility
K. Miller	Electrical Engineering
M. Mitchell	Management Oversight
D. Nguyen	Electrical Engineering
E. Patel	Consultant
N. Patel	Electrical Engineering
S. Pope	Consultant
J. Richmond	Region I Inspections
B. Rogers	Scoping & Screening
W. Ruland	Management Oversight
S. Sakai	Scoping & Screening
E. Sayoc	Project Manager
A. Sheikh	Civil & Structural Engineering
R. Sun	Audit Team Member
C. Sydnor	Vessels & Internals Integrity
J. Tsao	Flaw Evaluations & Welding
G. Wilson	Management Oversight
J. Woodfield	Consultant
E. Wong	Chemical Engineering
D. Wrona	Management Oversight
Z. Xi	Civil & Structural Engineering
Dr. C. Yang	Mechanical Engineering
O. Yee	Mechanical Engineering

APPENDIX
D

APPENDIX D

REFERENCES

Item Number	Reference
1	10 CFR Part 50, "Domestic Licensing of Production and Utilization Facilities."
2	10 CFR Part 51," ENVIRONMENTAL PROTECTION REGULATIONS FOR DOMESTIC LICENSING AND RELATED REGULATORY FUNCTIONS"
3	10 CFR Part 54, "Requirements for Renewal of Operating Licenses For Nuclear Power Plants."
4	10 CFR Part 100, "Reactor Site Criteria."
5	10 CFR 140.92,"Appendix B–Form of indemnity agreement with licensees furnishing insurance policies as proof of financial protection."
6	American Society of Mechanical Engineers (ASME) Boiler and Pressure Vessel Code Section III
7	ASME Code Class 1, 2 and 3
8	ASME Code Section III, Class 2
9	ASME Code Section III, Class 3, 1989 Edition
10	ASME Code Section III, Subsection NB
11	ASME Section XI Inservice Inspection, Subsections IWB, IWC, and IWD
12	ASME Code Section VIII, Division 1, and Section III, Subsection NC-3100/ ND-3000
13	ASME Code Section XI, Appendix G
14	ANSI B31.1-2007 standards, July 1 2007
15	Branch Technical Position (BTP) Auxiliary and Power Conversion Systems Branch (APCSB) 9.5-1, "Guidelines for Fire Protection for Nuclear Power Plants Docketed Prior to July 1, 1976"
16	BVPS License Renewal Application, August 28, 2007.
17	EPRI Technical Report Materials Reliability Program (MRP)-47, "Guidelines for Addressing Fatigue Environmental Effects in a License Renewal Application."
18	EPRI TR-103834-P1-2, "Effects of Moisture on the Life of Power Plant Cables,"Revision Final, August, 1994
19	EPRI TR-1 09619, "Guideline for the Management of Adverse Localized Equipment Environment," Revision Final, June, 1999
20	Federal Register (52 FR 32626) Leak-Before-Break Evaluation Procedures
21	Generic Safety Issue (GSI)-78, "Monitoring of Fatigue Transient Limits for Reactor Coolant System,"
22	GSI-166, Adequacy of Fatigue Life of Metal Components."
23	GSI-190, "Fatigue Evaluation of Metal Components for 60-Year Plant Life."
24	GSI-191, "Assessment of Debris Accumulation on PWR Sump Performance."
25	Generic Letter (GL) 80-113, "Control of Heavy Loads," December 22, 1980.
26	GL 88-05, "Boric Acid Corrosion of Carbon Steel Reactor Pressure Boundary Components in PWR Plants," March 17, 1988
27	GL 89-13, "Service Water System Problems Affecting Safety-Related Equipment," July 18, 1989.
28	GL 2004-02, "Potential Impact of Debris Blockage on Emergency Recirculation During Design Basis Accidents at Pressurized Water Reactors," September 13, 2004.
29	IEEE Standard 1205-2000, "IEEE Guide for Assessing, Monitoring, and Mitigating Aging Effects on Class 1 E Equipment Used in Nuclear Power Generating Stations,"Revision of IEEE Std. 1205-2000, March 30, 2000
30	Information Notice 98-26, Settlement Monitoring and Inspection of Plant Structures Affected by Degradation of Porous Concrete Subfoundations, July 24, 1998.
31	LR-ISG-19B, "Cracking of nickel-alloy components in the reactor coolant pressure boundary."
32	LR-ISG-2006-01, "Corrosion of the Mark I Steel Containment Drywell Shell."
33	NEI 94-01, Revision 1j, Industry Guideline for Implementing Performance-Based Option of 10 CFR Part 50, Appendix J, December 8, 2005.
34	NEI 95-10, Revision 6, "Industry Guideline for Implementing the Requirements of 10 CFR Part 54 – The License Renewal Rule," June 2005.

Item Number	Reference
35	NRC Bulletin 88-08, "Thermal Stresses in Piping Connected to Reactor Coolant Systems," June 22, 1988
36	NRC Bulletin No. 88-11, "Pressurizer Surge Line Thermal Stratification," December 20, 1988.
37	NRC Bulletin 2003-02, "Leakage from Reactor Pressure Vessel Lower Head Penetrations and Reactor Coolant Pressure Boundary Integrity," August 21, 2003
38	NRC Bulletin 2004-01, "Inspection of Alloy 82/182/600 Materials used in the Fabrication of Pressurizer Penetrations and Steam Piping Connections at Pressurized-Water Reactors," May 28, 2004
39	NRC Information Notice 2001-06, "Centrifugal Charging Pump Thrust Bearing Damage Not Detected Due to Inadequate Assessment of Oil analysis Results and Selection of Pump Surveillance Points," May 11, 2001.
40	NRC Order EA-03-009, "Issuance of First Revised Order (EA-03-009) Establishing Interim Inspection Requirements for Reactor Pressure Vessel Heads at PWRs," April 4, 2004.
41	NUREG-0612, "Control of Heavy Loads at Nuclear Power Plants," July 1980.
42	NUREG-0800, Revision 1, "Standard Review Plan for the Review of Safety Analysis Report of Nuclear Power Plants," Section 3.6.2 Branch Technical Position 3-1, July 1981.
43	NUREG-1057, Safety Evaluation Report Related to the Operation of Beaver Valley Power Station, Unit 2. Docket No. 50-412, Duquesne Light Company, et al., October 1985, Supplements 1 through 6.
44	NUREG-1057, Supplement No. 4, "Safety Evaluation Report Related to the Operation of Beaver Valley Power Station Unit 2," March 1987
45	NUREG-1437, "Generic Environmental Impact Statement for License Renewal of Nuclear Plants (GEIS)," Supplement 36, September 19,2008.
46	NUREG-1800, Revision 1, "Standard Review Plan for Review of License Renewal Applications for Nuclear Power Plants," September 2005.
47	NUREG-1801, Revision 1, "Generic Aging Lessons Learned (GALL) Report," September 2005.
48	NUREG/CR-5643, "Insights Gained From Aging Research," dated March, 1992
49	NUREG/CR-5704, "Effects of LWR Coolant Environments on Fatigue Design Curves of Austenitic Stainless Steels, "April 1999.
50	NUREG/CR-5999 Interim Fatigue Curves for Selected Nuclear Power Plant Components,"
51	NUREG/CR-6260, "Application of NUREG/CR-5999 Interim Fatigue Curves for Selected Nuclear Power Plant Components,"
52	NUREG/CR-6583, "Effects of LWR Coolant Environments on Fatigue Design Curves of Carbon and Low Alloy Steels,"
53	NUREG/CR-6934,"Fatigue Crack Flaw Tolerance in Nuclear Power Plant Piping A Basis for Improvements to ASME Code Section XI Appendix L," May 2007.
54	SAND96-0344, "Aging Management Guideline for Commercial Nuclear Power Plants - Electrical Cable and Terminations," September, 1996.
55	Regulatory Guide (RG) 1.43, "Control of Stainless Steel Weld Cladding of Low-Alloy Steel Components," May 1973
56	RG 1.163, "Performance-Based Containment Leakage-Test Program," September 1995.
57	RG 1.188 Revision1, "Standard Format and Content for Applications to Renew Nuclear Power Plant Operating Licenses" September 2005.
58	RG 1.190, "Calculational and Dosimetry Methods for Determining Pressure Vessel Neutron Fluence," March 2001.
59	RG 1.99, Revision 2, "Radiation Embrittlement of Reactor Vessel Materials,"
60	Westinghouse Commercial Power (WCAP)-16173-P
61	Westinghouse Commercial Atomic Power (WCAP)-11317, "Technical Justification for Eliminating Large Primary Loop Pipe Rupture as the Structural Design Basis for Beaver Valley Unit 1, March 1987" (including Supplements 1 and 2).
62	WCAP-11923, "Technical Justification for Eliminating Large Primary Loop Pipe Rupture as the Structural Design Basis for Beaver Valley Unit 2 After Reduction of Snubbers."
63	WCAP-12093, Evaluation of Thermal Stratification for the Beaver Valley Unit 2 Pressurizer Surge Line, Revision 0, including Supplements 1, 2, and 3.

Item Number	Reference
64	WCAP-12727, Evaluation of Thermal Stratification for the Beaver Valley Unit 1 Pressurizer Surge Line, Revision 0.
65	WCAP-15338-A, "A Review of Cracking Associated with Weld Deposited Cladding in Operating PWR Plants," October 2002
66	WCAP-15571, "Analysis of Capsule Y from First Energy Company Beaver Valley Unit 1 Reactor Vessel Radiation Surveillance Program."
67	WCAP-15446, "Analysis of Capsule 284° from the Florida Power & Light Company St. Lucie Unit 1 Reactor Vessel Radiation Surveillance Program," September 2000
68	WCAP-15677-NP, "Beaver Valley Unit 2 Heatup and Cooldown Limit Curves for Normal Operation," August 2001
69	WCAP-15571, Analysis of Capsule Y from First Energy Company Beaver Valley Unit 1 Reactor Vessel Radiation Surveillance Program, Revision 0.
70	WCAP-15571 Supplement 1, Analysis of Capsule Y from First Energy Company Beaver Valley Unit 1 Reactor Vessel Radiation Surveillance Program, June 2007.
72	WCAP-16527-NP, Analysis of Capsule X from First Energy Nuclear Operating Company Beaver Valley Unit 2 Reactor Vessel Radiation Surveillance Program, Revision 0.
73	WCAP-16527-NP Supplement 1, Analysis of Capsule X from First Energy Company Beaver Valley Unit 2 Reactor Vessel Radiation Surveillance Program, June 2007.

NRC FORM 335
(9-2004)
NRCMD 3.7

U.S. NUCLEAR REGULATORY COMMISSION

BIBLIOGRAPHIC DATA SHEET

(See instructions on the reverse)

1. REPORT NUMBER
(Assigned by NRC, Add Vol., Supp., Rev., and Addendum Numbers, if any.)

NUREG-1929, Vol. 2

2. TITLE AND SUBTITLE

Safety Evaluation Report Related to the License Renewal of Beaver Valley Power Station, Units 1 and 2

3. DATE REPORT PUBLISHED	
MONTH	YEAR
October	2009

4. FIN OR GRANT NUMBER

5. AUTHOR(S)

Kent L. Howard

6. TYPE OF REPORT

7. PERIOD COVERED *(Inclusive Dates)*

August 27, 2007 - June 4, 2009

8. PERFORMING ORGANIZATION - NAME AND ADDRESS *(If NRC, provide Division, Office or Region, U.S. Nuclear Regulatory Commission, and mailing address; if contractor, provide name and mailing address.)*

Division of License Renewal
Office of Nuclear Reactor Regulation
U.S. Nuclear Regulatory Commission
Washington, DC 20555-0001

9. SPONSORING ORGANIZATION - NAME AND ADDRESS *(If NRC, type "Same as above"; if contractor, provide NRC Division, Office or Region, U.S. Nuclear Regulatory Commission, and mailing address.)*

Same as above

10. SUPPLEMENTARY NOTES

11. ABSTRACT *(200 words or less)*

This safety evaluation report (SER) documents the technical review of the Beaver Valley Power Station (BVPS), Units 1 and 2, license renewal application (LRA) by the United States (US) Nuclear Regulatory Commission (NRC) staff (staff). By letter dated August 27, 2007, FirstEnergy Nuclear Operating Company (FENOC or the applicant) submitted the LRA in accordance with Title 10, Part 54, of the Code of Federal Regulations, "Requirements for Renewal of Operating Licenses for Nuclear Power Plants." FENOC requests renewal of the Units 1 and 2, operating licenses (Facility Operating License Numbers DPR-66 and NPF-73, respectively) for a period of 20 years beyond the current expirations at midnight January 29, 2016 for Unit 1, and midnight May 27, 2027, for Unit 2.

BVPS is located approximately 17 miles west of McCandless, PA. The NRC issued the construction permits for Unit 1 on June 26,1970, and on May 3, 1974, for Unit 2. The NRC issued the operating licenses for Unit 1 on July 2, 1976, and on August 14, 1987, for Unit 2. Westinghouse Electric supplied the nuclear steam supply system and Stone and Webster originally designed and constructed the balance of the plant. The licensed power output of each unit is 2900 megawatt thermal with a gross electrical output of approximately 972 megawatt electric.

This SER presents the status of the staff's review of information through June 04, 2009, the cutoff date for consideration in the SER.

12. KEY WORDS/DESCRIPTORS *(List words or phrases that will assist researchers in locating the report.)*

10 CFR Part 54, license renewal, Beaver Valley Power Station, scoping and screening, aging management, time-limited aging analysis, safety evaluation report, volume II

13. AVAILABILITY STATEMENT

unlimited

14. SECURITY CLASSIFICATION

(This Page)

unclassified

(This Report)

unclassified

15. NUMBER OF PAGES

16. PRICE

NRC FORM 335 (9-2004)

October 2009

Safety Evaluation Report Related to the License Renewal of
Beaver Valley Power Station, Units 1 and 2

NUREG-1929, Vol. 2

UNITED STATES
NUCLEAR REGULATORY COMMISSION
WASHINGTON, DC 20555-0001

OFFICIAL BUSINESS

www.ingramcontent.com/pod-product-compliance
Lightning Source LLC
Chambersburg PA
CBHW080246180526
45167CB00006B/2439